THE TIGER IS MY BROTHER

By Robert Elgin

THE TIGER IS MY BROTHER
MAN IN A CAGE

THE TIGER IS MY BROTHER

by

Robert Elgin

WILLIAM MORROW AND COMPANY, INC.
New York 1980

Library of Congress Cataloging in Publication Data

Elgin, Robert, 1921-
The tiger is my brother.

1. Elgin, Robert, 1921- 2. Des Moines Children's Zoo. 3. Zoologists–United States–Biography. I. Title.
QL31.E49A34 636.08'899'0924 79-25876
ISBN 0-688-03575-2

Book Design by Michael Mauceri

Printed in the United States of America

First Edition

1 2 3 4 5 6 7 8 9 10

For the members of my human family,
Jane, Rob, Joel, Becky, Michelle, and Bruce.
May they hunt with reverence and
defend the pride with love and courage.

Contents

Chapter *1*	The End of a Beautiful Friendship	11
Chapter *2*	Royal Lions and Sassy Snakes	19
Chapter *3*	Murderer's Row	38
Chapter *4*	A Desperate Gamble: I Hold the Losing Hand	42
Chapter *5*	The Timid Matador	54
Chapter *6*	Snakes Alive–And Loose	57
Chapter *7*	Chimps *Are* Smarter Than Some Humans	62
Chapter *8*	Becky Lion, Watchcat	95
Chapter *9*	Becky Lion Captures Santa Claus	106
Chapter *10*	The Zoo's TV Stars	110
Chapter *11*	The Mad Adventures of Brucie Tiger	113
Chapter *12*	There Is Only the Pack	123
Chapter *13*	Janie Lion: A Love Story	134
Chapter *14*	Affectionately Yours, the Big Cats	137
Chapter *15*	The Happy Christmas	147
Chapter *16*	Our Wolf Pack, Alpha and Omega	169
Chapter *17*	Hatari–Danger, Danger	177
Chapter *18*	Humans Bite Too	179
Chapter *19*	Leonardo, King of Lions	189
Chapter *20*	The Nature of the Beast	197
Chapter *21*	Zoos, a Last Hope	201
Chapter *22*	Free From Fear	211
Chapter *23*	A Happy Ending?	219

THE TIGER IS MY BROTHER

CHAPTER 1

The End of a Beautiful Friendship

It was Christmas morning, 1973. Skipper, the Des Moines Zoo's huge chimpanzee, was in a rage. The long black hair on his shoulders bristled in anger. He screamed—a long, shrill scream of defiance and hatred. Standing erect now, swaying ominously from side to side, he began his challenge dance, his feet pounding the heavy boards of his sleeping platform. Slowly at first, then faster, until the primitive rhythm became a booming crescendo of violence. Then suddenly, fangs bared, he charged across the length of the cage to where I was standing.

He hit the cage door with an impact that shook the heavy structure.

Skipper had long been one of my closest friends, but I decided immediately that I wouldn't go into the cage with him that morning. Something had decidedly upset him. He had "gone ape" in that wild, psycho way that only an adult male chimpanzee can. I kept a safe distance from the bars, fully respecting his long, powerful arms.

"What's the matter, old fellow?" I asked him softly. "Did you have a bad dream last night or something?"

The words seemed to calm him somewhat. He squatted down on the cage floor, "oooofed" at me in a friendly way, and reached for my hand as he always did. Still concerned, though, I kept my distance and decided to wait a few minutes before I approached him, permitting him to take my hand and begin his usual grooming ritual.

"Just a minute, Skip," I told him. "Wait until we're through cleaning your cage; then we'll play."

I turned my attention to my eighteen-year-old son Joel, who was sweeping the area behind the rear of the cage. "It's okay now, Joel," I yelled to him. "I'll keep Skipper amused over here while you sweep and hose down that part."

Joel began working closer to the cage and I relaxed for a moment and talked to Skipper.

"You don't care much for old Spotacus do you, Skip?" I said, referring to the spotted leopard who was pacing the floor of his cage just ten feet away across the aisle. We had placed the big cat there the day before with much regret, for we knew that such close proximity to his ancient enemy might anger the big chimp. We had no choice, however. In the Des Moines Zoo, space was always at a premium and the cage was the only empty one in the winter quarters.

I kept talking to the chimp while Joel worked, and Skipper seemed to relax and be his old self again. Once more he reached his huge hand out through the bars, obviously hoping I would give him my hand in return or bend down and permit him to groom my hair or my face.

Suddenly, two young lions began a desperate roaring battle in the cage behind me. Instinctively I turned to see what the commotion was all about.

My sudden motion threw my left hand forward. Skipper grabbed it instantly with both hands and drew it to his chest. In one terrible second I realized what he intended to do.

"Skipper's got me!" I screamed to Joel. I struggled frantically, trying to pull my hand away from the animal.

Then I heard a crunching noise as he bit off one of my fingers.

Joel was there instantly, pounding and poking the huge ape on the face, in his eyes, with the broom handle.

I screamed at the animal. "No bite–no bite," I shrieked, using the command signal we used on our affection-trained animals, forgetting that Skipper had never been taught to obey such a command.

Then I heard a second sickening crunch and cracking sound as the ape bit off another of my fingers.

Joel dropped his broom and dashed for the hose some twenty feet away across the room. Turning on the water, he raced back to the cage. I was seated on the floor now, feet braced against the cage, trying frantically with all my strength to break free from the chimp's powerful grip on my arm. A big chimp is as strong as ten men. I was completely helpless.

Once more the animal took my hand up to his mouth to bite me. His lips were pulled back, his huge tusks, bloody now, open and ready. Just then Joel hit him in the face with the full force of the water from the hose. Skipper hated water. He dropped my hand for just a moment. Just long enough for me to fall backward and away, to rise to my feet.

I stood there for a long minute, in a state of shock, just looking at the animal. His face, his arms, his chest were covered with blood—my blood. My two fingers lay on the cage floor, a foot-long tendon still attached to each. I had struggled so desperately with the chimp that the tendons had been pulled out of the entire length of my forearm before he had had time to sever them with his teeth.

I turned away, sickened at the sight. Holding my mutilated hand and squeezing it tightly with the other, I staggered with Joel's support from the building.

Someone on the zoo staff called an ambulance; another keeper rushed my wife, Jane, over to the zoo from our home a block away. Twenty minutes after the incident I was in the emergency room at the Des Moines Mercy Hospital. Fortunately an orthopedic surgeon, Dr. Marvin Dubansky, was in the hospital at the time and, at my doctor's request, he agreed to do what he could to repair my hand. He obviously didn't have much to work with. The chimp had bitten off my first and second fingers deep into the palm, broken and partially severed my ring finger, and badly lacerated my thumb.

"Count from one hundred backwards," the anesthesiologist instructed. I began the count as they injected the anesthetic; then I began to sense that warm, beautiful feeling of relaxation as all the fear and pain passed away.

Two hours later I awakened to find my hand encased in a big blood-soaked bandage. Dr. Dubansky, a huge, bearded giant of a man, was looking at the dressing. He smiled when he saw I was again conscious, reminding me of a big, happy bear. My wife was standing on the other side of the bed. There had been no hope of sewing the missing fingers back in place, the doctor explained. The bites had penetrated too deeply into the palm. He did believe, though, that I would regain the use of my third finger and thumb, but it would be months before they would be of much real use to me.

I glanced down at the bandage, obviously disturbed by his words.

"You're very lucky to be alive, you know," he suggested quietly.

I nodded in agreement. "Man, how well I know that," I replied. "And thanks very, very much for sewing me back together."

Dr. Dubansky pressed my shoulder reassuringly and left the room. He said he would return shortly.

Jane pulled up a chair close to the bed and sat down. The strain of the past few hours was evident on her face. She took my good hand in hers and squeezed it.

"Well, you made it again, Bob," she said with as much cheer as she could manage.

I took a deep breath, sighed, and nodded in agreement. How right she was. It had been a very close call, I had to admit.

"Two reporters from one of the news services are waiting in the hall. Do you want to talk to them for a minute or shall I tell them to come back tomorrow?" Jane asked.

"I think I can visit with them for a little while," I told her. "Ask them to come on in."

Jane walked to the door, opened it, and the reporters entered. They were both old friends and I was happy to see them. They expressed their concern and asked how I was feeling.

"Ugh!" I replied with as much emphasis as possible. They smiled. One took out a note pad and the other began putting a lens and other gadgets on his camera.

I began to explain what had happened. They, in turn, began taking notes and popping strobe shots at me. I attempted to look as happy as possible and continued talking. I gave them the full details, emphasizing that I had not been in the chimp's cage when he had grabbed me.

The reporter held up his pencil, interrupting me. "But wasn't the chimp a tame animal?" he asked.

"The big problem with chimps is that when they become thirteen years old, as Skipper is, the males often get very aggressive at times," I replied. "I felt that I was safe with Skipper, and Eddie Anderson certainly was. He raised Skipper from infancy and worked with him for eleven years before he gave him to the zoo. He could do anything with the chimp. Both of us were always careful, though, and when Skipper was having one of his temper tantrums we stayed out of his cage. That's exactly why I didn't go into his cage this morning. If I had, I certainly wouldn't be talking to you now."

The reporter shook his head in disbelief. "Why, then, Elgin, if the animal is so dangerous, do you keep him at the zoo?"

"Skipper is our biggest attraction," I replied. "He is almost as large as a female lowland gorilla, after all, and some of the crazy antics he performs make him a favorite with everyone. Actually, he isn't any more dangerous than most of the animals at the zoo. Beautiful little buck deer with big brown eyes, for instance, probably kill more zoo people than any other animal. Keeping the chimp, like everything else, is a calculated risk and we lost—or at least I did."

The reporter smiled and shook his head again. "The hospital staff must be getting to know you pretty well by now, Elgin," he said.

"As I recall, you've been bitten by a copperhead, two cobras, mauled by a leopard and a lion, and now this. And how about the twenty-five-foot python that almost strangled you?"

I managed to smile at his last remark. "That didn't happen at all and you know it. The python just got out of her cage onto the floor and the five policemen and I had a little trouble getting her back into her cage simply because she weighed two hundred and seventy-five pounds. Actually, she's a very nice snake and she never attempted to harm me."

"Elgin," the reporter asked—a little sharply, I felt—"just what are you trying to prove? After all, there's got to be something funny about all this derring-do bit of yours. Do you get your kicks out of being some hairy-type he-man?"

"Of course not," I retorted. "I'm no big animal kook and you know it. I'm too big a coward for anything like that. But if we don't do something very different out at that little zoo once in a while, the attendance drops to nothing, my staff gets cut, the animal collection dwindles, and pretty soon the weeds will simply cover the place. You know as well as I do that the city fathers are not going to continue using taxpayers' money to keep the zoo open unless people support it with their presence and partially with the admission revenue."

The reporter jotted down a sentence or two and looked up at me again. "Aren't you getting pretty old, at fifty-two, to be charming cobras, milking rattlesnakes, playing with big lions, and dragging elephants around for the kids to ride on?"

"Well, I'll have to admit the elephant gets pretty heavy when she leans on me around the corners," I said. "But I'll probably continue with the other activities if I'm able to after this. The little elephant rides and the reptile lectures are our only hope, believe me. And at least I'll keep on trying until we get a year-round exhibit building and open, natural-habitat areas for our animals that will give them enough room for exercise. Then I can step back, sit down with my feet on the desk, and be a good little, do-nothing-type zoo director."

The photographer nodded his head in silent agreement. The reporter asked me if I had anything else to add to the story.

"No," I replied, "you've got the whole bloody bit as far as I can recall. Do you have any more questions you'd like to ask?"

They both agreed they had plenty of material for their story. They wished me a speedy recovery, said good-bye, and left the room.

Jane turned the crank on my bed so I could sit up straighter. Then she sat down and just looked at me, long and hard. There was concern in her eyes, but there was also anger.

"Why," she demanded, "didn't you tell the reporters the truth?"

She didn't give me a chance to answer. "Why," she went on, "didn't you tell them that the federal safety inspectors made a tour of the zoo last summer and agreed with you that the chimp's cage must be covered with chain link to prevent him from reaching out and grabbing people? If the cage had been properly fixed you would never have gone into the building in the first place to protect Joel, and certainly the chimp would never have been able to grab you."

My wife is a fighter. And in a real sense she was right. This I knew, but I suddenly felt very much too old and too tired to care.

"What's the use?" I muttered. "Someone just forgot to tell someone else, that's all. There's no use in our crying over spilled milk. Besides, we've probably lost the battle anyway. I don't see how I'm going to handle snakes or the elephant with half a hand. And if we don't have our special activities, the zoo will probably, some day or other, just go down the drain like so many others have. Maybe I'd just better forget the whole thing and go back to doing public relations and free-lance writing."

I didn't really mean a word of it. And, as usual, my wife understood and said the right thing. "No," she said positively, "you can't do that. Those animals mean too much to you. You'd never be happy if you had to give up Brucie Tiger and Janie Lion and all the rest of them. Your hand is going to be all right. You've got to believe that."

Doctor Dubansky returned with a nurse just then. He picked up my hand and again examined the bandages. "How are you feeling?" he asked.

"My hand is beginning to hurt like blazes," I answered.

"Well, Bob," he said, "I can help you some in that respect, but this is going to hurt you for a long time. Be prepared for that."

"I'll practice a little self-hypnosis," I told him. "It's funny, though, Doctor. I don't remember feeling a thing while the chimp was biting them off."

"That's only natural," he explained. "A lot of my patients who've lost fingers in meat grinders or corn pickers tell me the same thing. You were probably in a state of emotional shock and so determined to get away from the animal that nothing else mattered."

"Doctor," my wife interjected, "he's concerned that he won't be able to handle animals with it in the future. Please tell us the truth. How much will he be able to use his hand once it has healed."

"Oh, I think you can relax about that," Doctor Dubansky replied. "It'll take months before he has full flexibility in the remaining fingers, but once that stage is passed he'll build up compensating strength and dexterity to make up for the loss of the two fingers. I don't see, really, why he can't do about everything he does now. But it will take time. Don't forget that."

His words made me feel immensely better. I breathed a very audible sigh of relief.

Dr. Dubansky grinned at me. "I'll give you some pain pills now and the nurse will give you something to make you sleep tonight. I'll see you in the morning when I make my rounds."

Dr. Dubansky and the nurse left the room. Jane came over to the bed and gave me a big hug. "Now," she asked, "don't you feel a lot better about your hand?"

I had to confess that I did.

"And don't forget," she said, "that we have everything in the world to be thankful for. You're still alive. And it's a wonderful Christmas, after all, because of that."

I had forgotten that it was Christmas. Then I remembered that first cobra bite years before, and almost dying, and the terrible, childish fear that kept saying I wouldn't be home for Christmas—ever.

Well, you are still here, old man, I told myself. Maimed some, maybe, and cut up a little, but you're still here.

Jane smiled at me. "I'm going home now," she said. "I'll bring the kids back with me later this afternoon, if you feel up to having visitors."

I assured her that I'd like nothing more than that. "Tell Rob, Joel, and Michelle I'm feeling fine. And tell Becky to be careful with her horse. Take some change out of my pocket and buy a candy bar for little Bruce, will you?"

Jane went to the door, paused for a moment, and then turned back to me. "Bob," she said, "Joel is afraid you feel he deserted you when he dropped the broom and ran to get the water hose."

I couldn't answer for a long moment. I was very close to tears. "He did the only thing that could have helped," I told her. "The broom handle was useless, really. He could have poked that psycho chimp for days and it wouldn't have discouraged the animal a bit. You tell

Joel he did just the right thing and that I'm very proud and grateful that he's that much of a man."

Jane walked back to my bed and kissed me good-bye. Just before she closed the door she turned and smiled. "I'll bring the whole pack in this afternoon," she said.

I had to laugh at her remark. Years ago, somewhat dismayed at a few of my zoo colleagues' overreaction to the anthropomorphisms some zoo visitors projected on the captive animals, I had formed the habit of doing quite the opposite. In my mind I created zoomorphisms, wherein I attributed animal qualities to humans. My family, in frequent flights of fantasy, became much like the "pack" in Kipling's *Jungle Book.* My wife, Jane, petite, with the exquisitely formed features of a beautiful cat's head, became my panther. First son, Rob, strong and wonderfully coordinated, became the jaguar of my pack. Second son, Joel, huge, with leonine features and a mop of unruly blond hair, was a lion. First daughter, Becky, lithe, long-legged, and as beautiful as the wolves she worked with, could only be my "wolf girl." Smaller daughter, Shelly, slender and athletic, with green, slanting eyes, was our tiger. Little Bruce, husky, big-chested, and blond, was my lion cub.

I lay there for a long time thinking about them and how lucky I was to be still with them. Then the pain in my hand began again and I began to feel down, way down, and discouraged.

What do you think you're doing, you old goat? I asked myself. That reporter may have been right. You've probably had it. For seven years you've been trying to change things and you're not one step closer than when you started. You just can't beat the establishment, old man, that's obvious. You're never going to have decent facilities for those animals, and you may as well admit it and quit fooling yourself.

Disgusted and discouraged, I lay there wondering how I could have possibly managed to get myself into such a mess. The harder I tried, the more trouble I seemed to create for myself.

And, inevitably, as I thought about the situation, my mind drifted back over the past seven years and I recalled some of the funny and scary things that had happened during my stumbling, bumbling attempts to build the Des Moines Children's Zoo into something like a real zoo.

CHAPTER 2

Royal Lions and Sassy Snakes

I remembered the happy days, years before, in the spring of 1968. It was during my second year as director of the Des Moines Children's Zoo. We became the proud owners of two lion cubs. This was our first big step toward becoming a real zoo–or so we thought at the time.

The Fort Des Moines Lions Club offered to donate the little animals, and I was delighted with the proposal. I immediately called my superiors in the Parks Department for permission to accept the gift.

"There's no money in the budget to feed lions, Bob," I was told. "I'm sorry, but you cannot accept the lion cubs." Then he added the clincher: "Besides, lions don't fit in with the image of a children's zoo."

I pleaded with him desperately. Somehow, I promised, we would find an inexpensive way of feeding them that the budget could afford.

"And please remember," I argued, "that many children's zoos do have lion cubs in their exhibits for children to pet."

"Well," he replied with much reluctance, "come on down to the office tomorrow morning and we'll talk about it."

I put the phone down and leaned back in my chair. Somehow it was always the same, I told myself. The darn budget always got in the way of everything we tried to do at the zoo. I remembered how difficult it had been during the previous year to purchase a bullhorn for group tours, the tranquilizer gun our veterinarian had said we must have, and the PA system we needed for our lecture programs. Kind friends had donated the money for everything when the budget had failed me.

I had been told, too, that snakes didn't fit in at all with the proper image of a children's zoo. I, in turn, argued that TV had made sophisticated zoo visitors out of most children and that they expected to see snakes, lions, tigers, and virtually everything else at the zoo. We convinced the Parks Department that snakes didn't cost much to feed and that building a snake box or two didn't involve much expense. Finally, after the intervention of one of the Parks board members, we were permitted to purchase a boa constrictor.

As I sat there that day, most discouraged, I decided to give my problem a name. Mr. VOB, Mr. Voice of the Budget, I determined to call it (meaning him, her, or them, of course, as members of the "establishment"). Somehow, it seemed quite appropriate.

I laughed aloud at myself. After all, Elgin, I mused, your budget and administrative problems are not terribly unique, you know.

I had just returned from a regional meeting of zoo directors a few days before. There had been only two real topics of conversation among the some fifty zoo directors who had attended: their animals and the impossible difficulties they were all having with the municipalities and zoo associations that administered their zoos. There had been only one exception. One of the members of the famous Topeka Zoo informed us, during one of the meetings, that he had brought along a very rara avis, and he wished to introduce him to the group. This man, he explained, was one of the few members of a parks department he had ever met or heard of who really tried to understand the many problems his zoo faced and had attempted to do something helpful about them. He introduced Dennis Showalter, director of the Topeka Parks Department. When Dennis stood up, every zoo director spontaneously rose to his feet and gave him a very long, appreciative ovation.

I spent the rest of the afternoon thinking up reasons why Mr. VOB should accept the little lions. One of my main arguments, I told myself, was the odd fact that the cubs would be the only animals in our children's zoo that a child could pet, or even get close to. The rest of the collection consisted of big, mean elk, buffalo, zebras, eagles, hawks, coyotes, and wolves (most unhappy and mean), exotic birds that no one could catch, some bad-tempered monkeys, a dangerous chimp, some baboons with huge canine teeth, and two big black bears. The name "Children's Zoo" was, in our case, a complete misnomer. We had nothing for the children to touch or play with, not even a petting area with little goats, rabbits, or lambs.

I had no real explanation for this. I was the zoo's second director, and why the first director (or whoever else was responsible) chose such a strange collection of animals for a children's zoo was beyond me.

This, I felt, was one strong argument that might impress Mr. VOB. Then I thought of another. The year before (1967) our attendance had dropped drastically. One of the reasons was probably that most of those who had attended the zoo in 1966 (the zoo's first year) did so out of curiosity. I was becoming quite aware by now that in the long run the attendance figures, and the resulting revenue from admissions, would possibly be the deciding factor in whether our zoo grew or declined. And certainly, as every zoo director knows, if a zoo declines in popularity, attendance, and revenue income, it will sooner or later be put to sleep by the city fathers or just wither on the vine and die.

Armed with these pertinent points and others in favor of little lions, I entered the Parks Department office the next morning. After a long discussion, a compromise solution was reached. We could accept the cubs, Mr. VOB agreed, but only for the exhibit season. After that we must sell them to another zoo. I accepted this arrangement, but deep down in my heart I resolved, somehow, someway, to make those lions so much a part of our zoo and the city of Des Moines that no one could possibly make us part with them at the end of the season.

We ordered the little animals and they arrived a few days later. They were very special—this the animal dealer had assured us. The parents of our cubs were owned by the emperor Haile Selassie of Ethiopia and were part of his private collection. The emperor had given some of these royal lions to the producer of the motion picture *Born Free*. We were told that the parents of our cubs had played the parts of little Elsa, her sisters, and other lions in the motion picture.

In later years we discovered something even more interesting and exciting about our lions. When our male lion cub grew to maturity and developed his mane, we were delighted to find he possessed a beautiful black mane that extended, thick and luxurious, back along the entire length of his abdomen and up his flanks. This unique characteristic suggested that our lion had a great deal of Atlas lion strain in his heritage. The Atlas lion once roamed the northern part of Africa and was certainly the most beautiful of the African lion species. It is now completely extinct in the wild and only a few specimens remain alive in zoos.

We placed our little lions in the main animal room of the winter quarters, in a small cage, just beside the small elephant's paddock. Actually we had no choice. The other winter animal building was also filled with small cages bursting with unhappy wolves, foxes, raccoons, and coyotes. There were no empty cages there.

The winter quarters were particularly distressing to me. I had flown hawks and falcons for thirty-three years—since I was twelve years old. My birds had always enjoyed the best of accommodations and, above all, they had flown free virtually every day. To see all the zoo's beautiful animals confined in cages so small they could hardly turn around was enough to make me most unhappy and virtually ill.

There were many times during my first year at the zoo that I sincerely regretted becoming a zoo director—simply because of the terrible facilities. I had been a free-lance writer and a public relations director for a savings and loan association when the position of zoo director opened up after the first director had been dismissed. My good friend Alan Blank, grandson of the man who had donated the money to build the zoo's exhibit area, had urged me to apply for the position. There was much I could do, he insisted, that would help the animals. In a burst of enthusiasm, I had filled out the necessary forms and, somewhat to my surprise, had been appointed director of the Des Moines Children's Zoo.

I've often wondered just how I got the job. Certainly my qualifications were quite limited. I had, over many years, established myself as something of an authority on the training of birds of prey. Indeed, I was probably the first person in history to fly goshawks free after only five or six days' training (it usually requires three or four weeks with these difficult birds), or to develop a technique for training a falcon to "ring up" thousands of feet in the air and "wait on" until the quarry was flushed beneath her. Neither of these accomplishments, though, really seemed to apply to anything in the zoo world that I could see.

Nor could I establish to my own satisfaction that my college background in liberal arts had helped my cause in any way. Though probably it didn't hurt it either, for I was to find out later that there is very little in any college major that really prepares one for working with exotic animals. Zoology is of some value in preparing the scientific names for zoo signs and lecturing on such dry subjects as taxonomy and the Aristotelian approach to evolution. Animal science courses and even a degree in veterinarian medicine are not in themselves sufficient, for exotic animals differ greatly from domestic animals. The public relations and administrative requirements of running a zoo are something very few zoology and veterinarian students ever acquire in college. There is, as I found, no substitute for experience in the zoo world. Some college fields may be of help, but actual experience is the main requisite.

In an effort to bolster my ego a bit, I would like to think that my long experience as a free-lance writer and public relations man had favorably influenced the Parks Department in their choice. At least I could write press releases, and, knowing most of the people in the local media, could, I told myself again and again, obtain some free advertising for the zoo. And, I rationalized, if I could keep such delicate creatures as raptors alive and in good health, and train them, I was somewhat adequately prepared to care for other exotic animals.

Actually, though, I'm forced, deep down, to admit to myself that I became director of the zoo because of Alan Blank's influence. Possibly it helped, too, that absolutely no one else applied for the job.

It was becoming quite apparent that our big dream of helping the animals was going to be a very difficult matter. I was beginning to be aware of that terrible little word *budget* and the problems it could create.

The building we placed our little lions in, for instance, was only sixty feet by sixty feet—TOTAL! The entire north wall consisted of a tiny elephant paddock, the cage housing our lions, and the exotic birds' cage. The center section was filled with two rows of small monkey cages, each just four feet high. Separated from these by a narrow aisle was a row of teeny-tiny cages, one on top of the other, containing small birds and mammals. The noise was simply a bedlam; the odors, since there was no ventilation system, were overpowering.

The building was certainly a most inappropriate place to house our two little royal lion cubs. Actually, considering the additional presence of millions of cockroaches, mice, and rats, the winter quarters was more of a catastrophe than a zoo. My only consolation that first year was that the animals would be moved to the summer exhibit area after a few weeks, and the facilities there were at least a degree more acceptable.

The little lions proved to be quite adaptable, thank heaven. They weighed just twenty-five pounds when we received them that spring. Both were blondes with fuzzy hairdos. They had large, inquisitive amber eyes, short, bowed legs, long tails, and teeth that could bite. The cubs had been weaned before we received them and were now on a diet of horsemeat and milk, but from the looks of our hands we were convinced that they preferred human meat. They were decidedly not friendly.

Obviously, this sort of relationship could not continue. Their snarls and bites would just get bigger as they grew larger—and they might simply eat us. Actually, though, as I had learned from working with birds of prey for so many years, their aggressiveness was simply a mask for the desperate fear they felt in their strange surroundings and the presence of even stranger two-legged, infinitely tall humans. I resolved to calm their wild little spirits and make friends with them somehow.

I chose junior keeper Gary Enfield to help me. Actually Gary looked very much like a lion himself, with his blond hair and classic, lionlike profile. I hoped this might help. Another consideration, too, was that the regular keepers, Earl Connett, Willard Collins, and Harold Wessell had assigned chores to do each day and lacked the time.

Since Gary and I had no idea of how to go about taming or training hostile little lions, I spent hours searching the library for books that might help in our endeavor. I came across two or three instructions on how to train a lion to sit on a pedestal, but since this was quite the opposite of what I wished to accomplish in our lion-human relationship these were of no help.

I did, however, stumble across a book by the renowned animal psychologist, Heini Hediger, entitled *Wild Animals in Captivity*. In it he described some of the psychological needs of animals in captivity and how they might best be met. A portion of one chapter in particular seemed to apply to our situation. Dr. Hediger stated in one paragraph:

> Compared with the wild state, tameness for the wild animal in captivity has only advantages. It should be strongly stressed that tame animals alone should be kept in ecological gardens. . . . Tameness in large animals is of great advantage when attending to them, changing their quarters, and especially when handling a number together. With animals that have no urge to escape there is no need to fear panic on such occasions, no attacks on their keepers, no attempts to break out or incidents of that sort. . . . To sum up the advantages of tameness: there are reasons for stressing the need for tameness in as many animals as possible in zoological gardens; tameness is attractive; tameness is healthy; tameness is expedient.

Hediger's statements certainly agreed with what I had learned about keeping falcons, hawks, and eagles in captivity. My only question was a matter of the proper technique. I knew, of course, that many zoos hand-raised lions when they were rejected by their mothers or when they wished to bottle-feed them and place the cubs in a children's contact area. Such animals, I had been told, were not reliable. Indeed, they seemed to become very unpredictable and dangerous as they grew older and bigger. Inasmuch as our two cubs were already past the bottle-feeding stage and quite large, even this approach didn't appear to hold much promise.

I recalled reading, years before, that the ancient Egyptians had managed to establish quite a remarkable relationship with lions. They used the big cats to hunt with and to protect the pharaoh when he ventured into battle. They even shaved the lions for some strange reason. This must have been quite a ticklish operation, to say the least.

A more recent example of an affirmative lion-human relationship is the story of Elsa in Joy Adamson's beautiful book *Born Free*. This, above all, proved to me that we might succeed in making our lion cubs a little happier about things than they presently appeared to be. If we couldn't give them the freedom that Joy Adamson had been able to in Elsa's Africa, we might at least give them freedom from fear. This seemed quite appropriate, for our little lions were, in a small sense, part of that wonderful story.

In the absence of any definite procedures to follow in taming and training the cubs, I decided finally to fall back on some old falconry techniques and simply "man" the lions, as a falconer would in the first phases of training his captive hawk.

This was the beginning of a behavioral technique we were later to call "affection-training." As we worked with our predator animals over the years, studied them, and gained new insights into their psychodynamics and behavior, we added new approaches and modified some of the old ones. Never, despite the fact that each animal is an individual with all the differences that go along with individuality, were we disappointed in our results. Our big cats, wolves, bears, coyotes, and other animals were taught to walk on a leash for exercise, enter the back portion of my station wagon, and play with us. We were able to establish with our animals a rather remarkable relationship based on mutual trust and love. We were even able to teach our big cats never to use their claws, and in that I feel we were quite unique. Most important, the animals accepted humans and confinement without the fear, resentment, and hyperaggressiveness that many captive animals evidence.

And this, after all, was our main purpose. In our small zoo, we must move animals back and forth between the winter quarters and the exhibit area frequently each year. We learned very soon that forcing the animals into transfer cages with prods, or virtually drowning them into submission with water from a hose, was certainly not a humane answer. Nor was the tranquilizer gun. During my first years at the zoo we tranquilized more animals than most zoos, simply because we had to move them so frequently, but I confess I dread the process. We once maimed a tiger by hitting a nerve in her thigh. This was an unavoidable accident, as the guns are not too accurate. Despite the latest drugs and techniques, the animal sometimes goes into convulsions or, as on two occasions at our zoo, tranquilized animals may simply go into shock and never recover. Other zoos have had much the same experience.

It is quite probable that I never would have affection-trained our big cats if our zoo had more adequate facilities, but, conditions being the way they were, affection-training seemed the only reasonable approach to our complex problems.

At first we merely sat in the cage with the two snarling cubs, hour after hour, day after day. We offered them meat from our hands—very carefully. We talked to them in soft, low tones and tried to mimic some of the vocal communication the cubs used between each other. Gradually they accepted us as fellow lions and permitted us to stroke their fuzzy little heads. Later we were able to pick them up, even to enter into their lion play. We continued the process of "manning" them slowly and carefully, so that they would not be frustrated and end up resenting us. And, when they did become a little too rough with their needlelike teeth, we taught them to respect a limitation. They had to learn some degree of respect for Gary and me, at least as much as they gave each other in their roughhouse wrestling, or sooner or later we human lions might provide dinner for our growing cats.

We decided that they could do everything with us they did with each other, except bite our bare hands or other parts of our tender human bodies. To prevent this we wore gloves, which we permitted them to pull on and chew to their heart's content. If they did bite us instead of the gloves, we rapped them on the nose and yelled, loudly and firmly, "No bite!"

Our lion cubs soon caught on. After all, they expected a certain amount of resentment from each other when the going got too rough in their play. After a few smacks on the nose they learned to respect us—and to love us because we loved them so much. They grew very rapidly into animals with wonderful dispositions—free-spirited and proud, yet gentle and considerate of the humans who worked with and around them.

Like the sixty other big cats we affection-trained in years to come, they grew larger than most zoo animals, adjusted well to their small (and atrocious) cages, were easy to move from one cage or one area to another, were healthier, did less fighting among themselves, and were easier to treat medically if they became ill or injured. Of no little importance, too, our affection-training saved members of the zoo staff from serious injury or death when human error placed some of our zoo people in dangerous confrontations with our big feline friends.

The most satisfying thing to me, however, was that they were free from fear. Like a well-manned falcon, they had lost the natural fear wild animals instinctively feel in the presence of humans. And they've remained that way in the intervening eight years—friendly and happy and relaxed with me and rest of the zoo staff. I was even somewhat pleased with myself. If I could not provide them with adequate open areas for exercise and had to confine them in cages, at least I had the satisfaction of knowing they were not continually frightened. They evidenced none of the hyperaggressiveness or the withdrawal symptoms one finds in the usual zoo animal.

We named the cubs Ramses and Nefer. Somehow this association with ancient Egyptian royalty suited them perfectly, for even as cubs they were something very regal and special. At least Gary Enfield and I felt that way about them.

As the latter part of May approached, we began moving the animals into the exhibit area. The zoo had been constructed in something of a long oval shape. If one were to view the zoo layout as an imaginary clock, the wolves, foxes, and coyotes were placed in four small cages that formed the northeastern quadrant of the exhibit's perimeter from one to two o'clock. The little elephant was herded into her paddock just west of them at twelve o'clock. Small burros and goats were exhibited in the northwestern quadrant, followed by the guanacos and very unhappy, frightened zebras. At nine o'clock stood the Birthday House and a McDonald's Barn for domestic animals. The southwestern quadrant included two deer paddocks and a big bear cage. A large aviary and smaller cages for birds of prey formed the southeastern quadrant. The concession stand and the administrative and ticket offices were located at three o'clock, completing the circle. An old mill building, a monkey island, and a small Noah's Ark, all surrounded by a moated water system, formed a long line down the middle of the circle.

No one seemed to be very happy about the arrangement. The wolves, foxes, and coyotes evidenced their fearful plight by endless pacing, which virtually wore a bloody path inside their cages. The zebras' enclosure was much too small and offered no concealment whatsoever. The monkeys on monkey island were either too hot or too cold. The bears were determined to tear their cage apart and go elsewhere. They kept the city workers busy rewelding the bars the big animals continually tore loose. Worst of all, to me at least, was the plight of the birds of prey. Their eight-by-eight-foot chain-link cages were impossibly small, and the unhappy birds kept dashing against the sides until their wings and tails were a mess of broken feathers and their beaks dripped blood.

Many times I looked enviously at the large park area just north of the zoo's parking lot. I had no difficulty at all dreaming of just how it might be developed into a year-round exhibit area. The space consisted of some twelve acres and was shaped in something of a natural amphitheater slope. In my mind I could see an exhibit building that would house our primates, the elephant, reptiles, and exotic birds. Following the circle around would be a large natural habitat, open areas for the zebras, a paddock for camels, and finally, some distant day, an indoor-outdoor display for giraffes.

Many of the animals, in my dream, would be permitted to roam free in the green grass during the summer months. A small tram on an asphalt strip could even carry our zoo visitors around the exhibits and among the harmless grazing animals and such big ground birds as cranes, ostriches, and rheas.

Then, above all, we would be a year-round zoo and people could visit the animals for twelve months rather than the three they were limited to by the present exhibit area.

Very important, too (as I was beginning to realize), a year-round facility would substantially increase our income and reduce our budget deficit. We would be less of a burden to the taxpayers of the city.

We had great hopes for the 1968 season. In addition to our two lion cubs, some new reptiles, and a small exhibit space for them, we had trained our little elephant and could now permit children to ride on her back. A number of other zoos were doing this and had experienced a great deal of success with elephant rides. We felt that the kids would enjoy the activity and we might increase our attendance as a result.

The new reptile exhibit helped greatly, too. The additional space permitted us to add a number of new species, so we were able to milk rattlesnakes and other pit vipers during our reptile lectures and show our zoo visitors the venom and poison fangs. The narration was given by a friend of the zoo who had a degree in herpetology. He managed to make the lecture interesting as well as educational. Our audiences were told how to identify the poisonous snakes of Iowa and how to treat a poisonous reptile bite properly. He also took great pains to point out the importance of reptiles in nature's scheme of things, and told the audience that harmless snakes should never be killed if one encountered them in the woods or backyard.

The reptile lecture was such an obvious success the Parks Department permitted us to purchase a superstar for the program. I found a friend who donated the money and, after a few phone calls to animal dealers, we managed to locate a big python. The dealer agreed to make immediate delivery.

We picked her up at the airport the next day and brought the crate to the zoo in the pickup truck. Excited beyond words, we pried open the box and carried the sack containing our reptile into an open area close to the administrative office. We were most anxious to see our first really big snake and also to give her a little exercise after her long journey.

Keeper Earl Connett, our zoomorphic grizzly bear, and I opened the sack while the other members of the zoo staff stood back, waiting expectantly. We placed the sack on the ground and the huge constrictor oozed out and slid with silent grandeur over the grass. We watched, fascinated by the rhythm of her movement and the awesome power in her thick, muscular body.

Penny Python (as we decided to call her later) was an eighteen-foot reticulated python. She had been captured in Asia a short time before she came to live at our zoo. The animal dealer we had obtained her from had worked with her, he said, and assured me she was perfectly tame.

She was a magnificent animal. Her skin was a rich, earthy brown tinged with dark geometric overlays. A strange iridescence in her coloring sparkled in the sunshine, giving her a radiance that was almost mystical. Spots of golden scales interlaced with the other colors, forming a bright reticulated pattern. This, of course, is why her species is called the reticulated python. She also had large, lovely pumpkin-yellow eyes.

The reticulated python is the largest officially recorded snake species in the world. Our new reptile was big by zoo standards, but she still had a long length to grow before she equaled the record size of thirty-three feet.

Still Penny was big enough to be dangerous. Just the year before we received her, a specimen of the same size had captured, constricted, and devoured a ten-year-old boy in India. There are a few—very few —other confirmed incidents of big pythons dining on humans over the past two centuries, but for the most part the giant constrictors confine their diets to rodents and other small mammals.

While we watched in happy amazement, our snake began to glide away from us and toward some bushes fifty feet away. Keeper Earl (Pappy) Connett was the first to see the problem.

"Mr. Elgin," he said, with a sly grin on his face, "if that python reaches those bushes she's going to coil up in them like a pretzel and you're going to have a heck of a time unwinding her and getting her loose."

"Mr. Pappy Connett," I replied, "it may well be that you're quite right. What do you suggest we should do about the matter?"

His grin grew even bigger. "Well, Mr. Elgin, since you're the director and the big snake handler around here, I suggest that you just go out and pull her back this direction before she gets too far away."

I hesitated. That snake looked very long, very round, and very strong. And I, after all, had never even touched a snake this big, let alone handled one. I looked at Mr. Connett. Mr. Connett looked back at me, significantly.

I took a long, deep breath and walked up behind her, most carefully. Placing my hands under the middle part of her long length I began to drag her back. Penny turned her big head, looked at me long and earnestly with her pumpkin eyes, but accepted my restraint without protest. Once I had her back to her starting position I placed her on the ground again. She hesitated for a moment and then started once again toward the bushes. And again I picked her up like a long log and brought her back to point one.

Evidently, I told myself, the dealer had been quite right. Penny was indeed very tame and didn't resent being handled. I became more confident. Once more she crept toward the security of the bushes and this time I walked around in front of her. Bending down, with nothing but pleased affection in my heart, I reached up and over the python to pet her on the head. This was a sad mistake.

I didn't begin to see that snake strike. She moved so rapidly it was impossible for me to follow her motion. She hit the palm of my outstretched hand with her many, many long teeth and the result was like being slashed with dozens of razor blades. It was only a warning strike. She made no attempt to grab me with her mouth, throw her coils about my body, and constrict me. After striking she merely settled back in a head-up, alert position, half-coiled and waiting for any further aggressive move on my part. I fooled her by leaping to my feet and fleeing to a safe distance.

My hand was bleeding profusely. I wasn't seriously injured, of course, but the slashes she had inflicted were long and deep. The problem was that I continued to bleed. While big, burly Pappy Connett placed a Ketch-all noose about her neck and the rest of the zoo crew secured her and loaded her into her exhibit cage, I was still bleeding. I bled, abundantly, long after they were through with their job. In fact, I oozed blood for more than an hour, despite the application of ice cubes, cold compresses, and bandages. Finally Connett simply covered my wounded paw with inches of bandages and the bleeding stopped.

"Thought that snake dealer said the critter was real tame," he said with a big grin.

"I think I've learned something about reptile dealers," I replied quite sadly.

I had also learned that you do not approach big pythons from the front, that you never really trust reticulated pythons. All the big ones I've known have been snappy and unpredictable, regardless of what the animal dealers might say about them.

The 1968 season was a happy one. Our attendance was up over the preceding year, and of course our revenue from admissions rose, too. This we could attribute to the special activities we had developed, such as the reptile lectures, the free elephant rides, and other events the children could safely participate in. On Adventure Day Sundays, we permitted them to pet domestic animals and the little lion cubs (under close supervision), ride the burros, hold a hooded falcon, pet a harmless snake, or ride in the zoo pickup on a safari trip through the Great Plains area. The children loved the special events, and the zoo staff thoroughly enjoyed doing them, too, although it did mean a great deal of extra work.

Most important, our newly formed Zoo Association was beginning to discuss the possibility of expanding the zoo into the adjacent park area, with a year-round exhibit building and big open exercise areas for the animals. I was pleased and happy with the zoo world, even somewhat pleased with myself.

Once again we herded and hauled our collection of animals back to their winter quarters and prepared to spend the next eight months in the dark, pest-ridden limbo. This time, however, when we passed the empty park area there was a faint glimmer of hope in my heart. My dream was getting closer to reality, I told myself. I was sure of it. If we could have just one more successful season next year, I felt certain the Zoo Association could obtain enough support and enough revenue, somehow, to give the animals the freedom they so greatly needed.

We had something else to be happy about, too. I was even rather smug about it. We still had our young lions. With the help of a veterinarian friend of mine, Dr. Jim Roush, we had concocted a special nutritious diet that was inexpensive as well. So inexpensive, in fact, that Mr. VOB of the Parks Department permitted us to keep our lions as permanent residents at the zoo.

We had a wonderful time with the two young animals that winter. By January they were a year old and Ramses weighed close to three hundred pounds, Nefer close to two hundred. Gary and I had trained them to walk on a leash, though they didn't always go in the direction we wished or stop when we requested. Then, of course, we continued in the direction the lions wished to go.

Gary Enfield often took them for rides in the zoo's pickup truck—one at a time, of course, for they were now so large that the two lions and Gary just didn't fit into the cab all at once. They thoroughly enjoyed riding and would stick their huge heads out of the window like family dogs.

Gary had Nefer out for a rather long time one evening and it worried me just a little. When he returned, I asked him where he had been and what, in dear heaven's name, he had been doing.

"I just took her up to McDonald's for a hamburger, Bob," he explained.

"You did what?" I asked, incredulous.

"Yes, it's the truth," he continued. "Nefer loves McDonald burgers and french fries. Ramses does, too, except for the french fries. Neither of them cares much for ketchup, though. They prefer them plain."

"Have you done this before?" I asked, still somewhat unbelieving.

"Oh, sure," he answered, "almost every night. Nefer was pretty hungry this evening and she ate six hamburgers. A pretty big crowd gathered around the truck to watch her eat. They asked a lot of questions. That's why I'm so late."

"Who pays for the hamburgers?" I asked, thinking of his modest salary as a junior keeper.

"Oh, I do," Gary said. "After all, they need a little special treat once in a while, don't you think? They're pretty special lions."

I nodded in agreement. "I can't argue with that, but you'd better make sure you keep enough money to buy yourself something to eat, too, you know. Those two big characters may just eat you out of house and home."

Gary promised he wouldn't overlook that little detail.

Since then we've raised thirty lion cubs in our home, and I've concluded that one eminent zoologist is quite wrong in his assumption that one of the main distinctions between man and the lower animals is the animals' preference for raw meat. All of our home-raised lions have eaten cooked meat with relish. Some of them even preferred it to raw meat such as hamburger or horsemeat. The lion's greatest gastronomical delight, we've found, though, is undoubtedly tuna casserole.

Another popular misconception is that lions (and tigers, leopards, and other cats) don't like or can't tolerate cold weather simply because they come from tropical climates. Ramses and Nefer loved the snow and we took them out for daily exercise that winter regardless of how cold the weather might be. The lions' four-wheeled traction, plus their huge floppy paws, gave them a distinct advantage over the two-footed humans at the end of the leashes. Gary and I often found ourselves being pulled along in the snow, head first, like human toboggans. This always delighted the big cats. Once they were aware we were down and incapacitated they simply pounced on us and drove us more deeply into the snow.

Gary was able to cope with the problem better than I was. He, after all, was just eighteen; I was a more venerable and fragile forty-six years of age.

Nefer, for instance, would grab me by a coat sleeve or pant leg and drag me here and there. Tiring of the game, she rested, panting and heaving, by sitting on me. Struggle as hard as I could, it was impossible to extricate myself from beneath her big body.

"Help, Gary, hurry and help," I yelled on many occasions. "Get this big character off me before I suffocate."

Gary would rush to my rescue and allow me to rise to a more dignified position by pulling Nefer off my kicking, squirming body by her tail. By that time both lions were as big as most full-grown African lions ever get. Yet despite their size and incredible strength, neither Gary nor I ever suffered anything more than an occasional scratch. We had been quite successful in the upbringing and training of our big, fuzzy friends and the lions knew exactly how much pressure they were exerting with their fangs and never harmed us. Like most predators, they touch and sense their world with their teeth and they can be infinitely delicate or devastating with their huge canines.

We had discovered, too, that a lion uses its claws in very much the same manner we humans use our hands and fingers. Essentially they are used to grasp with rather than slash. In the wild, the cat actually catches quarry with its claws and uses them to pull the victim to its mouth where the long canine teeth become the killing instruments. This little insight, and some careful training, permitted Gary and me to play vigorously with the cats without the danger of being clawed or even badly scratched. If the going got too rough occasionally, simply because of their huge size and weight, we merely snarled at them in lion language or shouted the command signal, "No bite!" in good old plain English. They understood this perfectly and always obeyed.

Not all of the animals were as congenial and cooperative as our two lion cubs, though. One copperhead was decidedly unfriendly. He took advantage of a small error on my part and I learned a lesson I'll never forget.

I was transferring some reptiles we had loaned the Iowa State Conservation Commission for their state fair exhibit back to our own exhibit cages at the zoo. Instead of dropping the ornery little reptile as I should have, I foolishly allowed sentiment (I've always been quite fond of copperheads, really) to prevail and gently placed the snake on the floor of the exhibit cage, relinquished my hold on the back of his neck, and then began to pull my hand out of the way. Mr. Copperhead simply turned his head and struck me with one fang before I could move even an inch.

I closed the door of the display cage and, rather stunned by his intemperate action, I looked at the little hole in my right index finger. An instant later one very long, fiery stab of pain throbbed through that poor portion of my anatomy and I suddenly realized that I had been envenomed by the copperhead.

Ten minutes later one of the zoo staff had me in the office of Dr. Walter Eidbo just five blocks away. By this time my finger was almost black, terribly swollen, and my whole hand was becoming giant-size. Dr. Eidbo, an old friend, was a zoomorphic elephant if ever I had seen one and, happily, was just as patient and gentle as one of those great pachyderms. He studied my sad-looking hand, frowned, then began to read the directions that accompanied the commercial antivenin for copperhead bites I had brought with me.

"I'm afraid I'm pretty allergic to horse serum antivenins, Doctor," I told him. "I was bitten by a rabid dog when I was a kid and I've been hypersensitive to horse serum ever since."

He put down the written material and looked at me thoughtfully. "Well then, Robert," he suggested, "let's just give you a sensitivity test with the material the manufacturers have provided and see what we come up with."

Dr. Eidbo broke the little vial containing the serum and, after obtaining the suggested quantity, injected it into the inner part of my left forearm. Almost immediately a large weal began to form and slowly began to spread. A few minutes later my stomach erupted with a bad case of very itchy hives.

Dr. Eidbo came to a rapid conclusion. "I don't believe," he stated, "that we should even think of injecting much of this stuff, unless you feel that a copperhead bite is very dangerous. You're apparently so allergic to the horse serum antivenin that I'm afraid we'll be risking anaphylactic shock if we give you any measurable amount of it."

I hastily assured him that to the best of my knowledge no healthy adult had ever died of a copperhead bite.

"Then," the doctor stated, reaching for another syringe, a small vial of something or other, and an ugly-looking scalpel, "that means that the proper treatment will be to incise the bite wound and apply suction."

I looked at him very unhappily. "It already hurts very, very much, Doctor. I'm sure that cutting into it will make it hurt even more."

He nodded in agreement and sighed. "There's no other way, Robert. Perhaps if I inject a little novocaine first it may help."

I yelped as he began injecting the novocaine around the area of the bite. After waiting a few minutes for the stuff to take effect, he picked up the big scalpel and began incising the area.

He made one incision and looked up at me. "Did that hurt much?" he asked.

If I hadn't been hurting so much, almost shaking with pain, I might have laughed. As it was, I almost cried. "Doctor," I implored him, "rather than do that again, please, please, just cut off the finger, won't you?"

He laughed (a little evilly, I thought) and handed me a towel to wipe the perspiration from my forehead. "Oh," he exclaimed, "I think we've done all we need to do there. We'll apply a little suction and probably elicit a good deal of toxin with lymph." He examined my hand and shook his head in obvious dismay. "I'm afraid, though," he continued, "that we'll have to incise the back of your hand to eliminate some of the swelling there. The skin is beginning to crack."

I groaned aloud at the thought. The doctor smiled again and with infinite care and expertise proceeded to carve three long lines, each one-eighth of an inch deep, across the back of my hand. I gasped, yelped, and said some very naughty words under my breath.

Finished at long last with this beautiful bit of surgery, he slapped me in the hospital for four days. There they wrapped my awful paw in big wet towels and doped me with every pain-killer known to man. Nothing helped. I soaked innumerable towels with gallons of bloody lymph, but that awful venom kept crawling up until my arm was twice its normal size and black, blue, green, yellow, and red all over. In a grotesque way it was quite beautiful. The pain wasn't, though. No way! During my stay in the hospital and for a long month afterward, I felt as though my index finger and hand had been squeezed to a pulp in a vise, that my arm had been pounded continuously for all that time by a sledge hammer.

Fortunately, though, very little of the venom affected me systemically. I became a little nauseated a few times, but, thanks to the incising and suction treatment Dr. Eidbo had applied, I experienced no real problems.

Free from that distress and any worry that I might become a fatality, I could concentrate on the pain alone. I found that almost unbearable. Once my wife turned the fan on me because the bedroom was warm and I almost screamed with pain when the air struck my swollen arm.

Looking back on it, I'm still as amazed at the sheer power of the venom as I was then. Based upon our countless venom milkings at the zoo, I'm certain that a medium-sized copperhead could never eject more than half a cubic centimeter of venom with both fangs. Since the copperhead that did me in struck with only one fang, I'm sure that I could not have been injected with more than one-fourth a cc, approximating some twenty-five milligrams of venom. This really isn't much venom and certainly the copperhead's venom is probably one of the weaker toxins in the pit viper's family.

How well I remember hoping at that time that I would never be bitten by a bigger, more toxic snake than that little monster.

CHAPTER 3

Murderer's Row

The budget, that big black cloud which hangs over every zoo, large or small, appeared again over ours early in 1969. And again it threatened our growth and very existence because, in essence, it curtailed the normal expansion that was necessary for our zoo's welfare.

A sports car club whose members were interested in the zoo wished to purchase a tiger cub for the collection.

Mr. VOB refused the donation: "I'm sorry, Bob," he said, "there's no money in the budget to feed such an animal."

Again I pleaded with him, just as I had about the lions. "Sir," I insisted, "the collection is so small at our little zoo that we simply have to add something to it once in a while if we're going to continue to interest the zoo visitors and keep them coming through the gates."

I paused for a deep breath. "After all, you know yourself that the kids and their parents were fascinated by the baby lions last year. I'm certain they will find a tiger cub just as appealing. Probably the newspaper will take a picture of the little animal and, once the word gets around, we'll have a bigger attendance and that will increase our admission revenue."

I was learning. After a long discussion Mr. VOB gave in and permitted us to accept the little tiger on a tentative basis. Like the lions before him, he must not eat too much.

I wasn't concerned about the cost of feeding our new tiger baby, but I did utter a short, sincere prayer to the heavens that attendance would be good that year. I was, by now, finally aware that our zoo's existence depended upon a budget computed on money derived from the property taxes in our city and that this source of revenue was about reaching its limit. Our little zoo must pay at least a reasonable share of its expenses or we would suffer—it was as simple as that.

We had great plans for the 1969 season. We now had both lions and a tiger and we had also enlarged our reptile collection. We could purchase reptiles without too much difficulty (from Mr. VOB) because they were not expensive, did not require expensive caging, and, of course, were inexpensive to feed. And, like all zoos that had a reptile exhibit, we had found that they were immensely popular. No one likes snakes at all, it seems. They hardly have a friend in the world, but people certainly do visit the reptile exhibits in zoos in very large numbers. Possibly we humans are just fascinated by monsters.

Our first new shipment of reptiles arrived from Thailand just before the start of the season. I still shudder when I recall some of the characters we had selected to fill "murderer's row" in the poisonous section of the exhibit. In addition to our copperheads and rattlesnakes, we now had such distinguished killers as cobras, blue kraits, banded kraits, Russell's vipers, gaboon vipers, African puff adders, and last, but not least, one big fifteen-foot king cobra.

This collection of killers gave me something to think about. What, I asked myself, would you do, old man, if even one of the smaller cobras were to bite you? Maybe the copperheads and the rattlers wouldn't kill a healthy adult; maybe the kraits, the adders, even the king cobra won't pose much of a problem because they're pretty slow in some ways, but what about the cobras?

That was my big concern. We had a number of Asiatic cobras and, since cobras are very vivacious and can move about with lightning speed in their cages, there was a possibility of danger every time we opened their cage doors to feed or water them. And, while we had antivenin on hand, it was of no use to me if I should be bitten. Like the copperhead antivenin I had been so allergic to when bitten a year before, the cobra antivenin was also derived from horse serum. Injecting it might be even more immediately disastrous than the venom itself.

There was only one other possibility I could think of. Like everyone else who has done research on reptiles, I had heard of the famous Bill Haast and the Miami Serpentarium. Mr. Haast had been the first man in history to immunize himself successfully against the venom of cobras. I gave the matter much thought and determined to bother him with my peculiar problem.

Bill was most gracious and most patient. "Yes, Bob," he replied, after I had explained my problem over the phone. "I think it's possible for you to develop enough self-immunity to protect yourself, but it's a long and painful process."

I hesitated for a long moment. "Well," I finally managed to stammer, "I don't see any other solution. The cobras are our star attraction, because they stand up and hood so nicely, and I feel like I should have some sort of protection, even if it is painful. Can you tell me about the procedure?"

"Oh, sure," he replied, "there's nothing secret about that. I'll mail you the full details and a gram of cobra venom if you wish to purchase it. You'll have to be very, very careful, though, and not rush things. If you do, you'll possibly become allergic to the venom itself. If that happens, you have no choice. You'll have to get rid of the cobras."

I assured him that I would follow his instructions to the minute, and asked him to forward a gram of venom at once, via air mail. I was anxious to get started, I explained.

I heard him chuckle over the phone. "I'll get the venom out within an hour, I promise, Bob. Just take it easy and slow, though. You may find that this hurts a bit."

I received the venom the next day and, with the help of two good friends, Dr. Earl Redfield and microbiologist Don Elefson, I began the process of making myself immune to the common members of the cobra family.

Dr. Redfield gave me my first injection of venom as soon as Don Elefson had filtered it, cultured it, and pronounced it free of bacteria. The first dose consisted of just one ten-thousandth of a milligram.

Moments later the area around the injection began to get red. I think my face, on the other hand, became quite pale.

"Does it hurt?" Dr. Redfield asked, regarding me speculatively.

"Like ten bee stings," I groaned in reply.

"Why don't you sit down on that chair for a few minutes before you go. Just to make sure you don't have any allergic reaction."

I waited for fifteen minutes. The pain didn't go away, but there were no indications of any allergic response. Dr. Redfield sent me on my way with instructions to return for another injection in three days.

The amount of that first dose was very small—not even enough to kill a mouse—but I imagined (in my hypochondriac mind) every possible reaction. As I drove home afterward, I was certain that my lips tingled, that my eyes weren't focusing properly, that I was nauseated, and that I found it difficult to breathe. None of these was really possible, of course. It was all my imagination.

The pain certainly wasn't, however. Even that tiny amount of venom hurt like mad. And as I rapidly progressed through the next three weeks, to larger and larger amounts of venom, the pain became greater and greater.

The minimum lethal dose of cobra venom that can cause a human fatality is around fifteen milligrams, according to most authorities. We passed this neat little figure with some trepidation on my part, but without any serious symptoms. Just for additional insurance we continued on past the "kill" mark until we reached a full twenty-five-milligram dose of the deadly stuff. My arm, after receiving this amount, was a dreadful-looking mess—red, black, and yellow and terribly swollen from my fingertips to my shoulder. In many ways the reaction was quite similar to the bite I had received from the copperhead. There was one happy difference, though. The pain, while quite agonizing, didn't last nearly as long.

I would never recommend the procedure to anyone unless the person must handle cobras frequently and is very allergic to the horse serum of commercial antivenin. The whole process is most painful, quite uncertain (since the ability to build antibodies against the small cobra neurotoxin antigen varies greatly among humans, apparently), expensive, and no one knows just what the effect of continued venom injections might be on the body over a period of years. As an example, I experienced several severe reactions when I attempted to switch (in later years) from cobra venom to tiger snake venom. The latter is so toxic, and apparently endures in the human system for so long a period, that my frequent small injections very possibly accumulated into something close to a lethal dose. No one can tell for certain what did happen, but until I do find out I'm abstaining from any and all tiger snake venom.

Shortly after we had reached the very high twenty-five-milligram mark, Dr. Redfield discontinued the weekly injections altogether. From then on I received a booster shot of five milligrams just once a month. I suffered a very sore arm for three or four days, but I confess I felt more secure when it came time to clean cobra cages every day.

CHAPTER 4

A Desperate Gamble: I Hold the Losing Hand

A week after the 1969 exhibit season opened, it began to rain. And it continued to rain and rain and rain for days and weeks. Our attendance and revenue dropped to almost nothing. The elephant paddock was a mire of mud and it stayed that way until July. All of our special activities were curtailed, too, though that didn't really matter since there were very few visitors to enjoy them anyway.

Whom the gods would destroy, it has been said, they first make mad. This the rain gods—or my peculiar circumstances—certainly succeeded in doing to me. I was frantic, almost beside myself with worry and frustration. In my mind a disastrous season would further postpone the dream I had entertained for three long years—the greatly needed, altogether necessary year-round building and the open spaces for the animals. This, and the thought of spending much more time in those vermin-infested, tiny, smelly winter quarters, made me ill.

I'm not a good worrier. I don't tolerate it well, and I cannot continue long in that unhappy state. Somehow, I just have to keep gnawing away at the problem and finally worry myself into some solution and action, but when circumstances, such as the rain, are really beyond one's control, that's often difficult to do. Other things bothered me, too. The mounting criticism about the small, inadequate caging of our animals in the exhibit area added greatly to my distress.

We now had a young leopard, Shannon, in addition to our lions and tiger. Like the other cats, she had been affection-trained and we took them all out for walks or a ride in the zoo truck almost every night so they would get their needed exercise and entertainment. The cats were quite content with this arrangement. Certainly they didn't manifest the obvious distress the wolves, coyotes, foxes, coons, and other animals had displayed when they occupied the same cages, but the public was still unhappy about the relatively small areas. And, to make matters worse, I agreed with them. No one in the world wished to get them into large open spaces where they could exercise more than I did. My desire and the constant criticism from the zoo visitors developed into something near a total obsession in my mind.

The rains finally stopped around the first of July and I determined to do something to make up for the lost time and low attendance. The fact that this something was more than slightly dangerous—that it scared the heck out of me—didn't stop me from planning and preparing for a special dramatic activity that would insure an immediate drastic rise in the zoo's attendance.

I had just finished reading Carl Kauffeld's book *Snakes: The Keeper and the Kept* and had been fascinated by his experiences with cobra charming. To my knowledge he was the only individual who had ever authentically attempted that gentle art in this country. The fact that it was so unique and exciting would, I was sure, create a great deal of interest. Our attendance and revenue, I told myself, even though far behind now, might even exceed last year's and the Zoo Association would have good reason to insist to the City Council, possible donors, and the public, that people really cared about our zoo. Then, I felt, someone would be able to stimulate support for our needed expansion and my poor animals would have some degree of freedom. It was as simple as that. I was sure of it.

It didn't take long to discover that things were not quite that simple. When idea meets reality nothing ever is, it seems. The cobras proved to be something of a problem. They apparently didn't wish to just sit there, all hooded and nice, and be charmed. Instead, they kept crawling out of their baskets. They did this very rapidly and, although I had faith in my newly developed immunity, I certainly felt uneasy about continually catching them by the tails with my bare hands and stuffing them back into the baskets. I did not then and do not now like to be bitten by a snake, however harmless, and I further confess that even a mouse bite disturbs me.

Cobra charmers in Asia have the advantage of picking and choosing from an unlimited number of species. If one doesn't perform well, they just go and find a more suitable specimen. Obviously I couldn't do this at the Des Moines Children's Zoo. We had only two big cobras to work with, since we had to keep Suzie Cobra, our black Malayan, in her exhibit cage for display.

The Indian cobra charmers had another advantage, as Mr. Kauffeld's book had pointed out. They could cheat. And cheat they almost always did, by sewing the cobra's lips together, by pulling its fangs out, or by milking it of its venom immediately before a performance. Since the first two solutions almost always resulted in the death of the snake from subsequent mouth infection, I certainly couldn't utilize these techniques. Milking the cobra was a possibility, but it had one big drawback. To milk a cobra of its venom you must handle it first. The technique involved is just as dangerous as the actual charming, possibly even more so. Then, too, it takes just a few drops of cobra venom to kill a human, and nobody I had ever read about could be absolutely sure of milking a cobra completely dry of its lethal poison. Indeed, I found a few cases on record where the charmer had attempted to milk his snake and discovered, to his dismay that he had not been able to extract all, or enough, of the potent stuff. I suppose the charmer reached this sad conclusion as he lay dying.

There are many ways one can make mistakes in handling poisonous snakes. The professional charmers in Asia have made their share and paid for it with their lives. Little did we know that we were to make a brand-new one (new at least in the eyes of medical science) and that the results would be almost the same.

We persevered in our attempts to get our kooky cobras, Huff and Puff, to act in a more dignified manner, to stand erect, hoods raised, while I blew away on the little flute. We experimented with different baskets, flutes, and motions. We practiced and practiced for two weeks.

The cobras began to improve—maybe they simply tired of having me chase them about the room. They began to accept the basket as their home and would stand in place and dance well as their confidence improved.

Once this portion of the cobra act was going smoothly, I decided to add some real excitement. I had discovered I could play a little game of "cobra tag" with the snakes once they were up and hooded. This consisted of first diverting their attention with a slight movement of one hand, then reaching over and tapping the tops of their heads with my other hand. It provoked some great action on the part of the cobras and it was really rather safe showmanship. The common Asian cobra is slow in striking from a hooded position and the human hand is capable of some of the fastest motion in the animal kingdom.

After two weeks of practice, I felt we were ready. Attired in a beautiful (and bizarre) green and gold satin snake-charmer costume, I carried my three little baskets out and we gave our first performance. The act was an instant success. The big cobras danced well and struck at the flute and my big nose with great vigor and ferocity.

Our narrator explained to the spectators what we were doing and told them a great deal about the value of snakes in nature's scheme of things. He pointed out that if it weren't for snakes we would probably starve, since they eat billions and billions of mice and rats all over the world. Man, he said, with all his technical know-how, would simply not be able to keep the rodents under control—they would eat all the grain in the fields and granaries, and man would probably starve.

"Snakes," he told our visitors, "do not have external ears. They cannot hear airborne sounds." He paused for a moment and snickered. "This is just as well," he went on, "for if the cobras could really hear Mr. Elgin's atrocious flute playing they would surely bite him out of pure horror and disgust. The cobras are not 'charmed' by the music, to say the least. They merely follow the motion of Elgin's flute and his hands. This is why they sway back and forth in a dancing motion."

Despite my narrator's caustic comments about my flute playing, it was a pretty neat bit of cobra charming. It was highly educational, too. We stressed that part of it and the crowds loved it. As I had hoped, the local TV stations and the newspaper gave us a lot of publicity and our attendance figures began to climb, rapidly. At last things were looking up.

Not everyone was enthusiastic, of course. My poor wife suffered terribly every Sunday afternoon. And frankly, I always had a very dry throat and excessive underarm perspiration in spite of using the best antiperspirants.

Many of my friends were alarmed and concerned. I was in the offices of an advertising agency one morning reading over some copy they were helping us prepare for the news media. The president of the firm entered the room.

"Good Lord, Elgin," he exclaimed on discovering my presence, "what gives with this cobra-charming bit? Have you gone out of your skull?"

I grinned at him. "No, John, nothing like that, believe me," I replied. "I'm still perfectly sane."

"Well, what's the reason, man?" he asked. "Those snakes kill people quite often, don't they?"

I took a deep, deep breath. I had explained the situation so many times I almost had my little piece memorized word for word. "John," I began, "for the first six weeks of this season we had virtually no attendance at our little zoo. This means we had a very little revenue coming in. This is not good. For unless we pay a certain percentage of the cost of our operation the city will cut off some of my meager staff. This, in turn, will necessitate dropping many of our special activities, which will, in turn, cause an even further drop in attendance. Eventually, the city will decide that people are just not interested in our little zoo or that it costs too much to operate. Then they will close it down. You know for yourself that some of the city fathers feel the zoo should be completely self-supporting and certainly it's far, far from being that. The cobra charming has increased our attendance and revenue greatly ever since we started it."

John shook his head. "What happens, Elgin, if you get bitten? An increase in revenue isn't going to help you in any way if you're six feet under."

"We've taken every precaution. I've been taking venom immunization shots for months now. We've done tests on agar plates, using my blood serum and cobra venom, and the lines of interaction indicate that I'm building antibodies which will protect me if I should be bitten."

"Antibodies or no antibodies, I sure as heck wouldn't play around with cobras. Nor would I break my back giving elephant rides, train big cats, or milk rattlesnakes like you're doing. It just isn't worth it," he stated emphatically.

"Well, I know you work pretty hard at your business here," I pointed out. "I'm just doing the same at the zoo. I believe that Des Moines needs and can afford a small, adequate zoo and that the animals deserve better facilities. It's as simple as that."

"Can't you think of some other way?"

"You know as well as I do that there's no money in the budget for advertising," I replied. "That's why I'm bugging you to help us get this stuff aired and printed in the news media, for free. Believe me, John, I find giving elephant rides, hour after hour, very tiring. I do not like milking rattlesnakes, and I certainly don't derive any pleasure out of charming cobras."

I paused for breath. "All I want to do, really, is to keep us open out there and build a broad base of interest and support for the zoo in the future. Now, you're in the advertising business, John; can you think of a better way?"

He looked at me for a long minute. "No, I guess not," he answered. "You're certainly getting plenty of free exposure in the newspapers and on radio and TV, if that's what it takes to get people out there. And you're keeping it on a pretty high plane, too, I guess. No, I don't suppose there's much else you can do without a hundred-thousand-dollar advertising budget."

We both smiled at that statement.

Neither John nor any of my other friendly critics could suggest an easier solution when it came right down to it. We continued with our cobra charming.

All of the local news people had given coverage to our snake-charming act except my good friends at UPI. Their new photographer, Jim Carr, had not had the opportunity to catch one of our Sunday performances. He felt it would make an interesting feature, so we arranged a special performance just for him.

I squeezed into my splendid Arabian Nights costume, carried out two baskets of cobras, and placed them in the center of the zoo's lecture area. The baskets contained our big cobras, Huff and Puff. I sat down beside them. Jim Carr set up his camera equipment and announced that he was ready.

Huff, a fiery six-footer, was our most reliable performer. I arranged his basket in the proper position before me. One of the zoo staff handed me the little flute and I trilled out a few miserable bars of cobra-charming music. The surrounding crowd, as usual, found this hard to bear. I persisted, however, content in the knowledge that the cobra couldn't hear a thing.

I opened the basket and Huff reared up majestically. Hood spread, hissing his hatred, he moved slowly back and forth, intent on biting anything that came his way. I moved the flute from one side to the other in front of the snake. His beady little eyes fixed on the flute and my hands, he followed gracefully, bending this way and that in perfect rhythm. But only for a moment. Huff was tired and lazy that day. He struck savagely at my fingers, missed, and then began to crawl from the basket.

Disgusted, I grabbed him by the tail and the middle of his body and attempted to stuff him back into the basket for a second try. Before I could close the lid he was out once more, crawling away, he hoped, to some quieter spot. Again I retrieved him by the tail; again I attempted to stuff him in the basket. The contrary cobra crept out for a third time.

The only solution, as I well knew, was to grab him by the back of the neck so that I could stuff him tail first into the basket and then pop his head in last. This way I could slam the lid down before he could get a start on me.

Rather than get to my feet and use a snake hook to pin his head, I chose to grab Mr. Huff by the neck with only my hand. Once more I caught him by the tail as he attempted to dart away. Hissing with real anger, he whipped up and around in an attempt to bite me. I moved my hand just enough to make him miss and, as he fell, I stabbed out with my left hand and caught him behind the head.

This was a mistake. I'm simply not left-handed. While I had done the trick many times before with my right hand, this time I was a fraction of an inch too low. Huff turned his head, opened his big mouth, and sank one sharp fang into the back of my hand.

He hit me with everything he had. I tore him loose instantly but the fang remained, sticking in my flesh. Blood streamed from the tiny wound, an indication that I had received a bad bite, probably in a vein.

I wasn't too concerned at the moment, though. I was sure that the immunity I had established would be sufficient to counteract the venom, and I had received a booster shot of venom only the day before. I thought this would also help. I couldn't have been more mistaken—no way!

Placing the cobra back in the basket, I apologized to photographer Jim Carr for the briefness of the performance. Jim said he had a suitable picture or two and insisted that I leave and see a doctor at once. I called my wife, told her what had happened, and assured her there was no reason to worry. Then one of the zoo staff drove me to Dr. Redfield's office several miles away.

The good doctor agreed that my immunity would probably counteract the venom, but he did insist that I enter the hospital for the night. Just for observation, he said, and to make sure I had the proper antibiotic protection against infection. This I really did not wish to do. I felt fine, I told him. I wished to go home. The doctor didn't argue with me. He simply went to the phone, called the hospital, and made arrangements for my arrival. I went with great reluctance.

I was placed, as a matter of routine, they told me, in the intensive care ward and I had a chance to visit with some of the nurses who had taken care of me when I was bitten by the copperhead. This time, though, I was thoroughly enjoying myself. The cobra bite hurt, but it was much less painful than the copperhead envenoming had been. I was relaxed, even feeling a sense of intoxication and euphoria. I ate some delicious chicken broth around 11:00 P.M. and settled down for the night.

For twelve hours after the bite I had hardly a symptom. This was most unusual. Cobra venom works very rapidly in most cases. The victim feels the effect, quite terribly, within a few hours, or even minutes if he receives a bad bite. Few persons who suffer a big injection of venom live more than eight or ten hours unless they get prompt and proper treatment with antivenin.

Me, I felt wonderful—for twelve hours. Then the lights went out, literally. My eyelids closed and I couldn't open them. My tongue became paralyzed. I lost control of my neck and leg muscles. The only parts of my body I could move were my hands and arms. I vomited blood.

Suddenly I couldn't breathe. My chest muscles were paralyzed. I was helpless, gasping for breath like a fish out of water. The nurses placed a respirator tube in my mouth to do my breathing for me. A priest came in and gave me the last rites of the Catholic Church.

I've read that death by cobra bite is an easy, happy way to leave this world. I didn't find it so. I struggled for air; inwardly I screamed for more air. Even with the respirator I wasn't obtaining nearly enough oxygen, or so it seemed to me. Inert, sightless, speechless, and unable to breathe sufficiently, I had the terrible feeling that my own body was pressing in on me and becoming my coffin.

I was certain that I was dying. Somehow a part of me seemed to separate itself from the rest and I visualized my oldest son, Robbie, standing beside my coffin. Then the strange thought occurred to me that I wouldn't be home for Christmas and somehow this made me angry. At once I felt an intense, utterly unreasonable hatred for everyone around me. They were going to keep on living; I was going to die.

I was conscious, acutely so, of everything that was happening. And I could use my hands. When Dr. Redfield and Don Elefson returned to the hospital (after an emergency call from the nurse), I asked them, using a child's slate to write on, to please give me the commercial antivenin we had on hand, even if I was allergic to it. To me it offered some degree of hope, and even death seemed preferable to this constant horror, this feeling of black suffocation.

They thought it over and came to a negative conclusion. Giving me the antivenin, they explained, would only substitute one set of problems for another. The allergic shock of the commercial antivenin could kill me in hours, or it could kill me instantly.

They decided, Dr. Redfield explained, to transfer me to the university hospital in Iowa City. There, he said, I'd have the constant attention of a number of specialists and also, as a last resort, Dr. Jan Smith, an authority on allergies, might be able to give me the commercial antivenin without its killing me.

Before they carried me to the waiting ambulance, I managed to write one more word on the little slate. HAAST, I wrote, hoping they'd understand. I felt the serum antibodies in Bill Haast's blood might be the only thing that could save me.

Dr. Redfield got the message. "Yes," he told me, "we'll get him for you. As soon as possible."

They carried me out to the ambulance and I didn't for a moment believe I'd ever see my good friends Don Elefson or Dr. Redfield again.

Dr. Marvin Silk, an anesthesiologist, rode in the back of the ambulance with me, while Jane, my wife, rode in front with the driver and his assistant. At 10:00 A.M. we started toward Iowa City, some 120 miles away.

About halfway there I began to drown. Cobra venom affects the salivary glands, and torrents of thick saliva poured down my throat into my paralyzed lungs. If my lungs filled up with saliva, even the respirator would be useless.

I'm certain that no doctor ever worked harder to save a fellow being than Dr. Silk did that day. He worked constantly and desperately with a small manual suction bulb to keep my lungs free, all the while trying to encourage me.

My wife did, too. She kept me going just as she had for hours. "Keep trying, Bob," she called back from the front, "please keep trying. You're going to make it. You're going to make it."

What happened next is a matter of documented fact. Fifteen miles from Iowa City, with the speedometer at ninety, a tire blew out. Somehow our driver managed to keep the vehicle on the road and right side up. He braked to a screeching halt and the two men got out and began to change the tire. Then the jack broke. Fortunately, a passing motorist stopped and kindly offered the use of his jack. The tire was changed and we started again toward Iowa City.

The portable respirator had been damaged, though, and refused to work. Now Doctor Silk had to pump air into my lungs with one rubber tube and try to suction the saliva from my throat with another. I can only be thankful that Dr. Silk was physically big and strong. He never gave up—I very nearly did.

I was conscious and very much aware of everything. My heart was pounding furiously as it struggled to keep me going. I lay there, inside myself, listening to it and wondering how much longer it could beat so hard and so fast.

We reached the hospital just before noon. The doctors immediately performed a tracheotomy and put me on a respirator in the critical ward. Dozens of doctors and nurses, it seemed, began working on me. They taped wires in place, inserted tubes everywhere, even placed a pacemaker in my heart as a precautionary measure against any cardiotoxin effects in the cobra venom.

In numb gratitude I could only wonder at how hard everyone was working to keep me alive, even though it seemed pretty hopeless to me. For three hours I remained in a critical but stable condition. The nurses constantly suctioned the saliva from my throat and lungs, but I still felt the respirator was not giving me half enough air.

That afternoon my wife brought me the first word of real hope.

"Bill Haast is coming," she told me. "He's leaving Miami in an Air Force jet and he'll be here in just a few hours."

Man, how that news helped. I felt more confident, more determined to keep fighting and stay alive. And somehow, the doctors, the nurses, and my wife managed to keep me going—with the respirator, suction, and their words of encouragement.

Bill Haast arrived early that evening. Then, as my wife told me later, the whole atmosphere changed. Bill is a small, dark, very intense man with an abundance of that stuff called charisma. He inspires confidence wherever he is without the slightest bit of affectation. Everyone felt better the moment he arrived. Certainly I did, as soon as I heard his voice across the room.

Jane led him to my bed. I managed a feeble handshake and tried to trace the word "Thanks" on the palm of his hand with my finger.

"You'll be all right, Bob," he said. "Just wait a few minutes until we get some of my blood into you." His voice was quiet, sincere, and very reassuring.

I heard him ask Jane how big the cobra was and how long ago I had been bitten. She told him and then he talked to the doctors about the blood work that had to be done. Since his blood was a different type from mine, they agreed they would have to separate the serum from the red cells. I would receive the serum and the all-important antibodies; Bill would be given his red cells back again. Then, if it became necessary to give me more blood later, he would have the strength to do so.

Bill left for the blood lab. The technicians drew a pint of his blood and separated the plasma, or serum, portion. Within thirty minutes after his arrival, Bill's human antivenin serum was slowly dripping into a vein in my arm. Then everyone waited—doctors, nurses, Bill, and my wife—hoping for some indication that Bill's blood would help me.

I heard them discussing why my own immunity had failed, at least in part. Bill and the doctors examined the question from every angle and concluded that the venom booster shot I had received the day before was probably the culprit. My antibodies had been preoccupied with the booster shot, and some of Huff's venom had managed to slip by and do its lethal work.

They all waited expectantly while Bill's serum dripped into my arm. But I disappointed them. No one noticed any change or improvement. I merely lay there like a big, gray, long-nosed old mouse, hardly twitching. But inside me something had happened! Suddenly I could breathe. I could breathe deeply on the respirator and I was getting enough air. It was a beautiful feeling; it was like music; it was living.

While they watched and waited, I drifted off into a wonderful relaxed, deep sleep. For hours I slept like a lazy, but now very much alive, old log.

Bill gave me more plasma for another transfusion in the morning. This time I responded more graciously. I surprised everyone. My diaphragm muscles began to function. I was breathing on my own. My droopy eyes still refused to open, but I managed to mumble a feeble word of thanks to Bill Haast.

"You're doing fine, Bob," he told me. "And don't worry about the thanks. Maybe someday you can do the same for me."

He had to leave, he said. The Air Force jet was waiting.

"Be careful with those cobras," he admonished warmly, "and come see me in Florida sometime." Bill Haast left the hospital and hurried to his plane. He had just saved his fifteenth fellow human from a terrible death by snakebite—me!

And, of course, it's the dream of my life to visit him at the Miami Serpentarium in Florida someday and actually see with my own eyes the man who saved my life.

From then on it was rather easy sailing. The paralysis soon disappeared. The doctors now tapped me with little rubber hammers here and there and shocked me slightly with electricity to see how my reflexes were. Often, too, someone would creep up and take samples of my blood to test. They were very curious and asked so many questions that the leader of my medical team roped me off and declared me out of bounds. I needed sleep, he said, not questions. It seemed a little unfair to the rest of the doctors. I was so grateful, I would have answered questions all day and all night.

I had plenty of time to think during the next two days and to count my blessings. I found I had many, many of those. With pencil and paper I concluded that over fifty persons—doctors, nurses, ambulance drivers, Air Force pilots and crewmen, the man who loaned us his tire jack, the highway patrolman who rushed Bill Haast from the airport, and Bill himself—had expended no little energy in keeping one middle-aged, balding, and sagging zoo director alive. All of them had shown a devotion and interest to my cause that went far beyond any financial reward they may have received.

There was another almost unbelievable part that touched me in a way I'll never forget. There were many critically ill patients in the intensive care ward while I was there. Many of them were dying of cancer or heart disease and they knew they had only a matter of hours to live. Still, somehow, they had heard of my less desperate plight and sent their wishes for my safe recovery.

One young doctor's wife was a patient in the ward. She had suffered a sudden, severe stroke. She was dying, but the doctor still found time to console and comfort my wife.

Their concern, everyone's concern, was so real, so spontaneous, and so unselfish, I could only marvel at it. And the single conclusion I came to was that deep down, beneath their fears, their hostilities, and self-interest, people are really beautiful and wonderful.

I am forever grateful for their kindness and the help they gave me. I'll just accept it and never forget any part of it in any way. And possibly some day, if I have the opportunity, I'll be able to help someone else with the same unselfishness and generosity they gave me.

Monday morning they released me from the hospital. Jane and I left the nurse's station, after expressing our heartfelt thanks, and stepped out into the main corridor.

My three eldest children were waiting there.

No man will ever see a more beautiful sight.

CHAPTER 5

The Timid Matador

A month later, though still terribly weak and anemic, I was back on the job full time. I had retired, at least for the season, from cobra charming. The reason was quite simple—I didn't trust my immunity. Every time I injected a live mouse with my supposedly protective blood serum along with the cobra venom, that mouse died more quickly than the control mouse which received the venom alone. It was possible that some tiny amount of venom remained in my system and I certainly didn't wish to be bitten by another cobra while I was in that condition. Why help the cobra along?

I began the slow process of rebuilding my immunity and, while I did this, I emceed the reptile lectures, and gave the elephant rides when we had our Adventure Days. I had to lean on the elephant most of the time, but since she usually leaned on me I felt this was only fair. All in all, I managed to make myself useful.

It soon became quite apparent that I was not to spend the rest of the season in complete boredom. Zoos just aren't built that way. At least ours isn't. In fact, ours, in at least one place, is so badly constructed that a certain amount of excitement is almost a mathematical certainty.

I glanced out of my office door one fine August day just in time to witness the weird spectacle of a junior boy keeper running madly around the outside fence of the Great Plains area. A few feet behind him, snorting and puffing, galloped our big bull buffalo. And just behind him came his wife, the girl buffalo. And just behind the girl buffalo another of our junior keepers was charging along, evidently chasing the entire entourage.

It was impossible to determine just what was taking place, but I was too horrified to attempt making sense out of the situation. Particularly since the whole kooky group was heading directly toward a crowd of zoo visitors.

I grabbed a broom and ran toward the spot of potential collision as fast as my legs could carry me.

I arrived a trifle late, but thankfully the human critter who was leading the melee had circled somewhat and managed to avoid our precious zoo visitors. They, in turn, realizing their plight, had retreated to a safer position behind a small fence.

When the frantic foursome circled the area a second time, I was in a better position. I took one hearty swing at the big buffalo and managed to divert his attention. He pulled up short, turned like a polo pony, and just stood there for one long moment looking at me with his little bloodshot eyes.

Very frankly, this bothered me. A certain evil glint in his stare clearly stated that my presence was not welcome. To be quite truthful, hoofed animals simply terrify me. The big cats, poisonous reptiles, wolves, and birds of prey I know and understand, but apparently the big hoofed animals don't like me even a little bit.

This particular buffalo had made life miserable for the poor members of our zoo staff on many occasions. Quite often we were forced to venture into the Great Plains area, either on foot or in the zoo pickup, for one reason or another. The pickup had numerous big dents in its body and many times we almost suffered the same fate. The city garage raised heck about the damage to the truck, but this was beyond our control. The buffalo evidently considered it a worthy rival.

I tried to be brave, though the bull looked like a mountain across the meager ten-foot distance that separated us. I yelled in desperation for one of the keepers to open the main gate to the area, and tried to remember what a properly trained professional matador would do under the circumstances.

With every muscle trembling, I slowly advanced toward the big black giant. I shoved my little broom at his nose. To my great surprise he retreated a foot or two. Probably my new approach to bullfighting simply astonished him.

I gathered a bit more courage and bopped him on the nose with the broom. It could hardly have done more than tickle him, but this time he backed off, slowly, ten feet more in the right direction—toward the open gate.

Pursuing my advantage I slapped him again and again with the business end of the broom and, when I had backed him into position just in front of the gate, I screamed at him with a tremendous sound. The huge old bull simply turned tail and ran through the gate, back into his proper domain.

Mrs. Buffalo wasn't much of a problem. As soon as she saw Mr. Buffalo back inside the exhibit area, she ran right in. Probably she was as frightened as I was.

We closed the gate on the big brutes and then attempted to reassure our frightened zoo visitors. They accepted the adventure in the best way, which was certainly more than I could do at the moment. Shaking with fear and excitement, I crept back to the safety of my office while the rest of the staff tried to repair the fence where the animals had escaped.

We had successfully closed the gate on the animals this time, but not on the problem. A week later they pushed beneath the fence again, and we had the whole dangerous procedure to go through once more.

I'm surprised, really, looking back on it, that animals ever bothered to stay inside the Great Plains area at all. The contractors who had erected the chain-link fence had put it on the outside of the metal posts, and it was a simple matter for the buffalo, yak, and Barbary sheep to push upward on the fence with their horns and crawl underneath it. To make matters even worse, there were many places where the rains had so eroded the soil around the fence that the smaller animals could just walk under and out without losing their dignity by so much as stooping.

We had called this dangerous situation to the attention of the Parks Department many times. For years the answer was always the same. No, Mr. VOB would exclaim, there is no money in the budget to repair the fence. You will have to wait until next year.

Certain repairs have since been made. The Parks Department placed long rods along the bottom of the fence and filled in the eroded areas. The animals cannot get out that way any longer. They're waiting eagerly, however, for they now have something going for them that promises even easier access to the outside world. The entire south side of the fence is almost ready to topple over. One good shove by the bull elk or buffalo will do the job and then every creature in the area will be running all over the neighborhood on the south side of Des Moines.

The bull buffalo escaped for yet a third time that season and we did the best thing possible. We sold him to an animal dealer and purchased a little baby boy buffalo to take his place. We were tired of chasing and being chased by the mean old bull and we hoped that by the time the new baby grew up to be equally dangerous, Mr. VOB would find the necessary funds in the budget to repair the fence completely. It's getting to be something of a close race, however, for the baby is now big and becoming increasingly bold.

CHAPTER 6

Snakes Alive—And Loose

The season closed and we again retired to the winter quarters. I managed to keep myself busy. Contrary to some people's thinking, the director of a small zoo doesn't just close up shop and go on a glamorous safari to Africa or Asia during the off months of winter. I had weekly reports to do, schedules to make out, and three or four nights every week I tried to promote the zoo's cause by giving lecture programs for service clubs. There were Zoo Association meetings to attend and Parks meetings as well. In addition to a zoo director's usual supervisory duties, I had to work daily both with our little elephant and with the lion cubs, take the bigger cats out for exercise, talk to salesmen, answer the office telephone, and perform errands.

Furthermore, I was personally responsible for the snakes and for the proper care and cultivation of the captive rat and mouse colony that made up our snakes' diet. Although it's only required once a week, feeding snakes is undoubtedly the most time-consuming job in a zoo. Not only have we many of them to feed, but snakes have terrible table manners and must be watched constantly.

For instance, Penny Python and Cleo, an eight-foot boa constrictor with iridescent geometric patterns and a glamorous red tail, share a cage. One day I gave a large rat to Penny and a smaller one to Cleo. I had turned away momentarily to bid good night to Suzie Cobra when I heard the sounds of furious struggling in Penny's and Cleo's cage. Having finished her own meal, Cleo had her jaws clamped onto the tail end of the larger rat, trying to take it away from Penny. My first impulse was to laugh. The idea of an eight-foot boa trying to do such a thing to a twenty-two-foot python was absurd, but then I realized what might happen and I almost cried. I grabbed each snake by the neck and frantically tried to separate them—to no avail. I yelled for Gary Enfield just as Cleo's beautiful head began to disappear down Penny's throat.

Grabbing a broom, I inserted the handle into Penny's mouth and tried to wedge it open. Gary arrived and worked the snake's jaws farther apart while I grabbed Cleo's neck and pulled. Fortunately, Cleo relinquished her hold on the prized rat and Penny relaxed her jaws long enough to allow me to extract Cleo from her throat.

Cleo's head was badly cut in three places and her skull was punctured just above one eye where one of Penny's teeth had penetrated. We feared the boa would either suffer brain damage and die or be blind in one eye, but snakes have remarkable recuperative powers. In a week Cleo was completely healed and as demure and charming as ever. She and Penny made up and are quite happy again together—at least until the next feeding time.

My main concern that winter, however, was rebuilding my antibody protection against cobra venom. Dr. Redfield and Don Elefson had completed some further research and decided to use a new adjuvant with the venom they injected into my arms. This material, mixed in equal volume with the venom, held the lethal toxin in my system for a longer period of time and gave my body a better chance to build up antibodies to counteract the very small neurotoxin molecule in the cobra venom.

Dr. Merlin Kaeberle, a microbiologist at Iowa State University, volunteered to help with our project. He gave us invaluable advice on immunity buildup and how to prevent me from becoming allergic to the venom itself.

Dr. Sherman Minton, the leading authority on elapid (the cobra family) venoms in this country, titrated my blood serum and concluded that I should be able to counteract some three hundred milligrams of cobra venom. Since the usual cobra bite results in an injection of approximately fifty milligrams of venom, I felt that I had ample protection against the bite of most Asian cobras.

We also took several units of my blood serum with the protective antibodies, as we've done every year since then, and Dr. Kaeberle kindly lyophilized (freeze-dried, like instant coffee) them. This process will preserve the serum and antibodies for a long, long time.

This has made it possible to store my human antivenin for my own protection, should I receive a massive injection from a cobra bite, and to easily transport it anywhere to another victim of a cobra bite who might also be allergic to the commercial cobra antivenin. Perhaps I may be able to save someone someday, just as Bill Haast saved me.

We had a very frightening experience that winter, just to keep us on our toes. Snakes are beyond question the most elusive escape artists nature ever invented. Despite the greatest precautions, it is almost a mathematical certainty that, at least over a period of years, one or more specimens will eventually get out of a cage. The Des Moines Zoo was no exception, of course, and one day our blue krait—a real gangster—slipped out of his box.

Junior Keeper Gary Enfield burst through the door of my little office. "Bob," he shouted, shaking with excitement, "I can't find the blue krait. It's not in its cage."

"Did you look in the den, Gary?" I asked, completely at a loss to explain how the reptile could have escaped from the snake box.

"It's not there either," he exclaimed. "I peeked through the air holes first and couldn't see it in either the main compartment or the den. Then I opened the top doors and couldn't find it. That snake just isn't in that cage!"

We didn't waste any more words. Both of us simply flew to the part of the building where the snakes were kept. We examined the snake box again. It was decidely empty. Somehow the reptile had done the impossible—as snakes often do. The krait had in some way managed to squeeze through one of the small air holes. It could not have happened, but it was apparently the only possible explanation.

We had real trouble on our hands. A blue krait is one of the deadliest snakes in the world. Drop for drop its venom is much more potent than a cobra's and, if a person is bitten, the victim has only a 50 percent chance of surviving, even if he receives the proper antivenin immediately.

Blue kraits are rather small reptiles, which made our problem all the more difficult since the reptile room was rather large and divided into two areas by a door. To complicate matters further, there was a large drain in the center of the floor of the area where the krait was kept that led directly into the main animal room beyond the wall. If the snake had gone down this drain he could now be anywhere in the entire winter quarters building. That meant, I fully realized, that it might take days, even weeks to find him. And there was always a possibility that we would never find him. That would be a catastrophe, I told myself, and probably call for a brand-new zoo director. After all, the keepers and casual workers would hardly be happy working in a large building with a myriad of drains and countless nooks and crannies that could easily conceal the deadly krait.

"Let's start moving some of these other snake boxes," I suggested. "But for heaven's sake be careful, Gary. Let's take them by the top and scoot them around. That way we'll be able to keep our hands up from the shelves where the krait may be coiled."

We shifted every snake box in the room, peering carefully behind each one of them. We found nothing. Gary and I looked at each other for a long, unhappy moment.

"I suppose we may as well start looking in the main room," he suggested.

"I suppose we might as well," I replied, in the same sad tone of voice that Gary had used.

We searched the big room just as diligently. We looked into every corner, behind the refrigerators, beneath the heavy tables, and around the other snake boxes we kept there. We combed the floor, inch by inch. Mr. Krait had apparently vanished into thin air.

Or, as Gary suggested: "Do you suppose he went down that drain, Bob?" He had voiced my deepest fear and I became almost physically ill at the thought.

At a loss for anything else to do, we again entered the small room and lifted the lid from the drain. Both Gary and I got down on our hands and knees and peered into the hole. There was nothing there that we could discover.

"Well, I suppose the critter may have gone on through to the main animal room," I said, disgusted and discouraged. "If he did, we're going to have one heck of a time finding Mr. Krait."

"Yep, that's for sure," Gary replied, "if we find him at all."

That statement didn't do much for my morale, which was low enough at the moment. "Should we start there now?" I asked him. "Or do you want to rest for a while?" I was not only dejected, I was tired. We had searched the rooms for over an hour, and moving refrigerators, tables, and snake boxes had been plain hard work.

"We'd better start looking in the animal room," Gary answered. "Who knows, we might be lucky enough to find him before the rest of the crew comes back from the summer area."

I nodded in agreement. Gary rose to his feet. I turned, still on my hands and knees and began to stand up as well. A short, sharp hiss stopped me and I froze with fear. There, inches away from my nose, lay the krait, coiled and waiting to strike. Throwing myself backward, I leaped out of striking distance and the reptile relaxed from his threatening stance.

"Oh, man," I managed to mutter, my voice shaking, "that really shook me. I don't think I'll ever be the same again."

Gary was as pale as a ghost. Everything had happened so quickly he hadn't been able even to shout a warning, though he had seen the snake at the same moment I had.

"Hand me that snake stick with the red handle, will you?"

Gary walked over and picked it up, handed it to me, and a moment later we had Mr. Krait on the stick. There was an empty cage close by and I dropped him into it. The cage was for a smaller snake and the air holes were correspondingly smaller. This time our deadly friend would stay put permanently—we hoped.

Actually if the situation hadn't been so very serious it would have been laughable. All the while we had been searching like fiends in every possible place, the krait had been hiding right under our noses—almost literally, as I had discovered. He had been coiled just behind the vertical two-by-four that framed the door. Gary and I must have walked by him, just inches away, many times.

We were fortunate. The krait was the only poisonous snake that has ever escaped in our winter-quarters buildings, but other non-poisonous reptiles have. In one way or another two harmless snakes have managed to sneak out of their boxes and have never been found again.

Quite often I hear stories or read articles about persons keeping highly poisonous snakes in their homes. I've been known to become quite violent on this subject. Poisonous snakes belong in their natural habitat or in zoos, but certainly not in homes. I know too many of my friends who have found their favorite boa constrictors in cupboards, ovens, walls, beneath refrigerators—or never at all.

Boa constrictors are, of course, relatively nice and, being non-poisonous, make rather desirable pets. If they escape from their cage in your home you usually have very little to worry about—unless a sudden reappearance might frighten a visiting friend into a catatonic state or a heart attack. If, however, a poisonous snake escapes in your dwelling, your problem is infinitely greater. Some poisonous reptiles can live without food for months and during this long, anxious period in your life you may well be in danger of being bitten. But it's unlikely that your house guests will be in much danger, for you'll probably wish to vacate your home yourself for an indefinite period.

CHAPTER 7

Chimps Are *Smarter Than Some Humans*

Our 1971 season was a big success. Again our elephant rides, reptile lectures, cobra charming, and snake cult pageants paid off, even though these activities entailed a great deal of extra work for the zoo staff. Our attendance increased and our revenue went up.

Another new attraction to our zoo family accounted for a part of our increased attendance that summer. This was the year Skipper Chimpanzee came to live with us, and this great animal became one of the most popular we've ever had.

Eddie Anderson, Skipper's owner and trainer, called me from his home in Grinnell, Iowa, to tell me he had a trained chimp he wished to donate to the zoo. The chimp had a repertoire of more than thirty tricks, he said. Would I consider driving to Grinnell and taking a look at him?

I gave the matter a moment's thought and told him I'd call him back later in the day. We had never used any of our animals in acts. It seemed to me somehow out of place to have our beautiful, affection-trained big cats sitting on pedestals and jumping through hoops, and such antics certainly didn't have a place in the behavior research we were pursuing. When we took them out of their cages each day for their walks, we did so because I felt they needed the exercise and the play. I confess we did take advantage of the situation, however. Whenever a sizable crowd of interested zoo visitors came to watch them in their exercise areas, we always used the opportunity to explain that they weren't *killers* at all, that they only killed to eat, and they were certainly capable of forming an affectionate relationship with human beings. The children in the audience liked to hear this, but we always warned them that big cats should never be kept in their homes. They should be kept where they could receive loving attention, but always be controlled at the same time.

Chimps are such clever animals, though, that I decided I really didn't feel any reluctance about having a chimp perform for our zoo visitors' enjoyment. I knew too that the St. Louis Zoo had had tremendous success with their trained-chimp act in the past.

Another factor that influenced me (and made me hurry to the phone that afternoon to accept Mr. Anderson's kind offer) was the loneliness of our young girl chimpanzee. She would need a husband some day and I hoped Skipper might be an appropriate suitor. I started off for Grinnell that very afternoon.

I was most surprised at Skipper's size when I first saw him in his cage at Eddie's home that day. I had never seen an adult male chimp any closer than across a zoo moat, and Skipper looked as large as a gorilla to my astonished eyes. Actually, I wasn't far wrong. Skipper weighed a little over 175 pounds at the time and a female lowland gorilla is often about that size.

"My gosh, he's big," I muttered, flabbergasted.

"Yes, he's pretty good size, but some of the males get even larger. Some of them weigh as much as two hundred pounds," Eddie replied. "You stay here for a minute, Bob," he continued, "and get acquainted with Skipper while I get his collar and leash."

He left the room where Skipper was housed in his big cage. Skipper "ooooofed" at me and patted the floor with one hand. It was clearly an invitation for me to sit down beside the cage. He then reached out through the bars with his huge hand (more than twice the size of mine) and I gave him my hand in return. Very gently he took my hand in his mouth and bit down gently with his awesome dental equipment. This, I learned later, is part of the chimpanzee greeting ceremony. Happily, I permitted him to do so without any reluctance or hesitation on my part.

The great ape then sat down on the floor of his cage and motioned for me to come closer to the bars. He then began the chimpanzee's social ritual of grooming me. Ever so carefully he examined my hair, my face, and my eyebrows with his long fingers. While I didn't know it at the time, I had been initiated and accepted into the great and honorable clan of the chimpanzee.

Eddie Anderson returned with Skipper's collar and leash. Skipper screamed in sheer delight when he saw the equipment. He knew he was going outside.

Ed opened the cage and placed the collar around the chimp's big neck. "Come on, old fellow," he said, "Bob is going to give you a new home in his zoo. You'll have a lot more room to exercise there and all the oranges and apples you can eat."

The two of them passed through the house and out toward the car. I was content to follow, but Skipper wouldn't have it that way. He stopped and beckoned for me to come closer and take his other hand.

The three of us then walked, hand in hand, over to the door of the car.

Skipper made it very plain that he wanted to sit next to me on the front seat. Eddie was relegated to a second-class position in the rear. If I had any question as to why Skipper insisted on this seating arrangement I was soon enlightened. When we were speeding at seventy miles per hour on the interstate back to Des Moines, Skipper suddenly decided he didn't trust my driving. Reaching over with one long arm, he easily took the wheel away from me and attempted to steer. I'm sure it was just because he was in the wrong position, but we very nearly went off the road and into a ditch.

Eddie yelled at him. "Stop that, Skipper," he shouted, reaching forward and pulling the chimp back to his side of the car. "You let Bob do the driving. He's a good driver and he'll get us there without your help."

Skipper settled back in his seat, probably more relaxed since Eddie had reassured him about my driving ability.

I had much to learn about chimps, especially big adult male chimps. Years later, my friend Jack Rhodin, the famous Ringling Brothers chimp trainer, told me that in Europe, where the circus audiences are more sophisticated than they are here, the big emotional high of the entire show is when the chimps are brought into the ring—not the lions and tigers. Even later, I was to learn just how true his statement was.

My education began, however, that very evening when we arrived at the zoo. There were still a few visitors remaining as the three of us left the car and walked over to the cage Skipper was to occupy. Skipper hung back, apparently wishing to greet a few of his new admirers. To my surprise Eddie refused to permit him to do that. Instead, he led him as rapidly as possible into his new cage.

"What's the hurry, Eddie?" I asked, curious. "Why didn't you want to introduce him to some of the visitors?"

"Skipper was beginning to get frustrated," Eddie explained once we were inside the cage. "He is very much like a human child in temperament, and he doesn't like to be frustrated."

One glance at the enormous muscles in the chimp's big arm convinced me that I certainly didn't want to be the one that ever frustrated him. Some authorities have estimated that a big chimp has strength equivalent to that of ten men his own size.

Eddie and I seated ourselves on Skipper's sleeping shelf while the chimp explored his cage. Then, as they always do just before sunset, the lions began to roar. Skipper had been captured in Africa just before he was three weeks old and probably had never heard a lion roar, but some deep instinct told him the lion's roar meant danger.

To my astonishment he began to dance, to sway from side to side, faster and faster until the sound of his feet slapping the hard concrete was like a machine gun firing. Then he wheeled and charged at me and began flailing at my legs with his long arms and huge hands.

Eddie leaped between us and seized Skipper's mouth with both hands.

"Get out of here, Bob!" he screamed at me. "Get out of here as fast as you can!"

I did so, rapidly, though I was still somewhat puzzled and, in my ignorance, not really alarmed.

Once I was outside, Eddie released the chimp. "Open the door for me," he shouted. Skipper was very angry by this time and was chasing Eddie around the cage. Eddie managed to sidestep him each time. Then Skipper made a determined lunge at him; Eddie was just able to avoid the charge and rush out the door of the cage.

"What in heaven's name happened to him?" I asked. "Did he go crazy or something?"

"I'm sure it was the lions that frightened him," Eddie answered, panting from exertion. "Let's get him something to eat. Maybe that will calm him down."

That was the only time I ever saw Eddie have the slightest difficulty with Skipper during the ensuing three years. Even in this instance he must have had some degree of control over the animal, for he was able to hold the great ape's muzzle closed.

This relationship between Eddie and Skipper was most remarkable, as I was to learn and fully appreciate later. Very few chimp trainers work with animals over six years of age, and those that do often beat their animals with clubs to command the respect so necessary to the trainer's survival. Sometimes, too, the exceptional chimp that performs past sexual maturity has had his teeth removed or is controlled by a specially designed collar that permits the trainer to give the animal a strong electrical shock if he becomes threatening. Other trainers use whips to command the animal's respect and obedience.

Eddie's relationship with Skipper was quite different. He never found it necessary to use any restraint, any force or discipline with Skipper, excepting that first night. He commanded the animal with love and love alone. He gave many, many performances with Skipper during the next three years and he never had to resort to a whip, a club, or an electric collar. Skipper trusted Eddie and never betrayed him after that one first lion roar.

"Skipper likes and accepts you, Bob," Eddie explained after our hasty exit from the cage. "If he didn't, I'd never leave him here. But be careful. Never go into his cage and work with him when he's having a temper tantrum."

Since I had just experienced a temper tantrum, I had no doubt that I'd be able to recognize any future ones and I certainly had no desire to be in the same cage with Skipper when he had another.

"You work with him by yourself," Eddie continued. "Don't let anyone else even near his cage. Don't let any of your keepers ever take his hand if he offers it to them, under any circumstances."

He paused for a moment. "We had an unfortunate experience a few weeks ago," he said. "A man tried to take a cookie away from Skipper that a woman had given him while they were in my home. Skipper bit off four of his fingers."

I was simply stunned. After all, the chimp could easily have bitten off my fingers and half my face just a few hours before. Yet I was to find from experience that Eddie was right—mostly. Skipper had accepted me as a friend and a blood brother. Eddie was also right in warning me not to permit any other person ever to get close to his cage. To do so would simply be a matter of suicide—or murder.

I went home that night still in a state of shock. That weird dance Skipper had performed in the cage kept bugging me. The incredible energy and determination of the ritual and the attack revealed a primal world of such ferocity and frenzy that I found it most disturbing. It took me back to a world so savage, so violent, so devoid of reason, that it made me deeply afraid.

Skipper's performances at the zoo that season became even more popular than our reptile lectures. He could do anything and everything. He rode his bicycle (he could also ride a motorcycle but there wasn't enough room in his cage for that), was an expert on roller skates, walked a tightrope, brushed his teeth with a toothbrush, ate with fork and spoon, carried his dishes to the proper place for pickup, washed his face and hands, took a bath, did backward and forward somersaults, stood on his head, walked on his hands, opened his own flip-top pop cans, combed his hair, used his own toilet, groomed Eddie and me, clapped his hands, wore human clothing, and helped dress himself. Occasionally he threw things at people he didn't like.

In total, as Eddie had promised, he could do over thirty tricks and in my mind this made him probably the most accomplished performing chimp in the country.

I say this because the chimps that are used in circus performances usually do only two or three tricks in their acts. Actually, since there are always a number of chimps in a circus act and each does his own thing and another follows, rapid-fire, appearances are deceiving. Skipper's performances equaled those of a whole group of circus chimps and he did them all by himself.

A few years before humans began to teach sign language to chimps and converse with them in this manner, Skipper was teaching us humans to communicate with him in his own chimp language. We had no difficulty at all understanding him, what he wanted, what he disliked, even what he preferred. His vocal hoots, his screams of approval, his wide-mouthed smile, his body positions, his challenge dances, and his bull-like charges at the cage bars were clear expressions of preverbal behavior. We learned to understand and comply.

He had one uncanny ability that I've never been able to understand. He could tell instantly when Eddie Anderson arrived on the zoo premises in his car and would break into screams of utter delight. Possibly Skipper had some strange sixth sense, because there was no way he could hear the car, which Eddie always parked some distance from the exhibit area. Furthermore, he could not see or hear Eddie as he came down the service road on foot and entered the zoo's kitchen to prepare Skipper's food, because the kitchen building cut off the chimp's view of anyone approaching from that direction. Nor could he have smelled Eddie because the wind was almost always from the wrong direction.

Somehow, though, he always knew when Eddie was anywhere in the zoo area. And whenever we heard those high, shrill screams of pure joy we were certain that Eddie would soon appear. Never once was the chimp wrong.

Every spring when the zoo opened and every fall when it closed, Eddie and I drove Skipper back and forth between his winter quarters and the exhibit cage. On a few occasions we even permitted him to steer the car. He was a pretty fair driver, too. I'll have to concede that he was a much better roller skater, bicyclist, and motorcycle rider than I'll ever be.

One fall when we returned him to the winter building, we were given a demonstration of the loyalty and concern chimpanzees extend to their fellow chimps and even their human friends. We were temporarily boarding a young male cougar for the city (the animal control people had picked him up because his owner permitted the animal to run loose all over the neighborhood) because there was no place in the city's animal shelter to keep the big cat.

Concerned that Skipper might not like his new visitor, we placed a big tarp over the side of the cat's cage before we led Skipper into the building through the side door.

We should have known better. Skipper was suspicious the moment we approached the doorway. Eddie took his hand and attempted to lead him to his own cage. Halfway across the room Skipper stopped, convinced now that something bad was certainly behind the tarp. The cougar snarled. Skipper instantly started back toward the door, dragging 230-pound Eddie Anderson with him.

Suddenly the chimp stopped and turned. He had forgotten his other friend, Bob, who at the moment was still standing inside the room beside the chimp cage. Skipper gestured frantically with his free hand for me to come with him. He had no intention of leaving any of his clan in such a dangerous place.

He then held out his hand, motioning for me to take it. Hurry, he was obviously saying, and leave here before something eats us all. I walked over, took his hand and the three of us went out the door, sat down, and had a long conference about the matter. We "oofed" and hooted at each other, then great ape Bob (me) decided that if he were to go into the room he might be able to coax Skipper into the cage with a couple of bananas.

This approach almost worked. I lured Skipper back to the doorway and part way across the room when he suddenly decided I was being very foolish about the matter. Rushing up to me, despite his obvious fear of the thing that lurked behind the tarp, he seized my hand and literally dragged me from the room and back again outside the building.

Once again the three of us convened in conference. Eddie Anderson came up with a suggestion this time. "Why not just take the tarp off the cage," he said, "and let Skipper see the cougar behind it. I think he's afraid because he's not sure just what is there, and I doubt that he's much afraid of a cougar anyhow."

I nodded in agreement. Skipper "oof-oofed" in chimp language. Eddie entered the room and unhooked the canvas from the cage. Skipper, peering around the corner, watched him closely as he did so.

The cougar, now suddenly revealed, snarled savagely at Eddie. This was too much for Skipper Chimpanzee. Now that he saw the cougar for what he was, his courage returned. Further, his friend Eddie was apparently in danger.

The coarse black hair on Skipper's shoulders was instantly erect, bristling with hate. He screamed in anger, charged, and hit the cougar's cage like a bull.

The cougar was terrified. Skipper was probably the first chimp the cat had ever seen, after all, and he obviously didn't want a better look. The cougar darted to the rear of the adjoining shift cage where he curled up and hid.

Satisfied that he now had the situation under control, Skipper took Eddie's hand, came back to the door for me, and the three of us walked triumphantly into his roomy cage.

Eddie brought in Skipper's toys: his bowling ball (to roll, balance, and walk on), bicycle, rubber ball, table, and chairs. We opened three cans of pop and celebrated Skipper's return to his winter home.

Eddie and I left a short time later. As we closed the door, Skipper was resting on his sleeping ledge, eating a banana, and watching the late movie on his TV set.

Skipper was a splendid creature, beautifully proportioned, clean and healthy, and wonderfully intelligent. Researchers state that chimps have intelligence equivalent to that of a six- or seven-year-old child, but I'm certain that Skipper was more advanced than that. I believe he understood everything Eddie Anderson said to him. The German philosopher, Oswald Spengler, wrote that the eyes and the expression therein are the one great distinction between man and beast. Spengler apparently never had the opportunity to watch a chimp's actions or to look into the eyes of a big mature specimen. Skipper's eyes, and the obvious intelligence manifested in them, were as human as any I've ever seen: contemplative, alert, even penetrating at times.

Richard Leakey, the noted anthropologist, has recently found hominid remains in Africa that indicate that man is not a direct descendant of the chimpanzee. It's not difficult to believe, however, that they are our close cousins. A chimp's smile, while rather a frightening expression because of his big front teeth and canines, is very close to the human smile. The chimp's "oof-oof," rising to a crescendo of screaming delight, is certainly one of the first belly laughs in the primate family.

There is a pleasure syndrome in chimps, and in some other animals, that warrants further research. When a human friend approaches or, particularly, when food is presented, the chimp often manifests a spontaneous sexual response. While man has apparently lost most of this instinctive reaction, there may still be some truth to the old adage that the best way to a man's heart is through his stomach.

Chimps and humans have one other behavior trait in common. Unlike other animals, they often become hysterical without the slightest apparent reason. The big cats, even the reptiles, are relatively reliable compared to the hair-trigger insanity of the upper primates. With big cats and reptiles the defense reaction is always predicated on direct threat and the dynamics of attack or escape. With chimps and humans there seems to be more of a conscious ego involved. If, for instance, I lacked the time to stop and talk and pet one of my lions or tigers as I passed the cage, the animal would not be greatly upset by my neglect. It would be just as friendly and eager to see me an hour later. Not so with Skipper Chimpanzee. To pass his cage and not stop for grooming and a little talk evoked shrieks of rage from the huge ape. He actually felt rejection, I'm sure, just as a human would. And it required a cautious approach and maybe a special treat when I did return and visit him, even a short time later.

Virtually all the members of the animal world engage in combat with others of their species. There is, however, something in the emotional makeup of most of them that provides limits and safeguards to these intraspecies conflicts. When big cats, reptiles, hoofed animals, and birds engage in combat with their own kind, they attempt to establish dominance in one way or another. When the beaten animal submits or flees, the victor is content and makes no further attempt to destroy his adversary.

Chimpanzees and humans, however, seem to have left behind these instinctive checks and limits in their emotional dynamics. Only man and the chimp, apparently, consciously seek out or ambush members of their own species with the intent to capture and kill.

Skipper Chimpanzee often has a look of desperate loneliness in his eyes. He undoubtedly has the same emotional need for companionship that a human has. Without it he feels rejected, even ostracized. And this is the one need that humans, too, cannot endure too long or too deeply.

Both humans and chimps are highly developed family, or tribal, animals and we humans only fight and destroy our own kind when we feel (for the slightest reason sometimes) that we've been rejected by our fellow beings and cast out from the group. This is the cold, ice-blue world of the psychopath, the suicide, and the killer. This is the world more dreaded than death itself.

Probably Skipper felt much this way that Christmas morning in 1973, for I had been neglecting him somewhat. I just hadn't had the time to visit him for a week or so. He still recognized and knew me, of course, but very possibly the boredom of his cage existence and the absence of his one real human friend resulted in that terrible outburst of hostility.

And Skipper tried to kill me.

The only two trained animals that have ever seriously injured me have been raised and trained by other persons. Probably it's a matter of basic identification and bonding, for while Skipper liked and trusted me there was apparently a limit to our relationship. On the other hand, I'm quite certain that Skipper would never have attacked Eddie Anderson for the same reasons he did me. I believe many exotic animals form really close bonds with only one Alpha, or master.

Our research and affection-training experiments have included a large number of real problem children of the animal world. We've trained a number of cobras, rattlesnakes, and copperheads and handled them frequently with complete confidence that we would not be bitten. Our affection-training has also included many of the big cats, wolves and other canids, primates, raptors, elephants, and very difficult camels. I've never been more than scratched by any of them, though I must confess I did experience a couple of unpleasant moments.

I like to think that these animals loved me, respected me, or at least were certain that I would not harm them, and consequently did not harm me. Somehow, though, I failed to get the message across to Skipper Chimpanzee, and I completely failed to communicate with a big, mean old leopard we imported into the zoo to become Shannon Leopard's husband.

The cat belonged to an animal dealer who had loaned him to us for breeding purposes. The leopard, he assured me, was very tame, but since many big male cats often chew up or even kill the females when they're first introduced, the animal dealer suggested we tranquilize both his leopard and Shannon and then place them in the same cage. When they awakened, he explained, they would be drowsy for hours and during this happy period could become more closely acquainted without the danger of our lovely Shannon becoming a casualty.

I have no idea of whether or not his technique works. We've never had such a problem with our affection-trained cats, of course, and on this occasion our experiment didn't prove anything because I goofed.

Tom Thompson, a burly, big-chested, bullnecked man who is the zoo's zoomorphic Cape buffalo, and I gave the boy leopard an oral sedative in a pan of horsemeat which he devoured with obvious relish. Within a half hour he was virtually out on his feet. He was unable to walk and was having a great deal of difficulty standing. From time to time he flopped down on his stomach and stared at us blankly. It was apparent that he would soon be sedated.

Then, to my horror, I discovered that I had left his big pan of water in his cage. Realizing that if the leopard attempted to drink in his present state of extreme intoxication he would undoubtedly drown, I opened the cage door, grabbed the pan, and attempted to pull it out of the cage.

Cats, as any veterinarian will tell you, often react quite strangely to sedatives. If something arouses their attention they frequently become very active instead of just relaxing and drowsing off as they should. Mr. Leopard did just that. As I struggled to remove the heavy pan of water, the leopard revived with a vengeance. Snarling horribly, he sprang and seized me by the arm before I could move an inch.

In an instant he flipped me over and began tearing and clawing at my back. I didn't even pretend to be brave. I screeched for help. And when the cat began to work up toward my shoulders and neck I yelled even louder for Tom to please, somehow, do something. For in the position the cat had me, I certainly wasn't able to do much to help myself.

Tom solved the problem in his own original way. Leaping up on the guard fence, he jumped down onto the neck of the leopard with his 225 pounds and his heavy boots, and the leopard tore himself away from me. Freed, I rolled away as fast as possible, and Tom swung the cage door against the leopard, pushing him back into the cage.

After locking the cage, Tom turned his attention to me. Helping me to my feet, he hoisted me over to the zoo truck a few feet away and began driving rapidly toward the office of the zoo man's best friend, good, graying Dr. Eidbo.

Once there they cut away the tattered, bloody remnants of what had been my shirt and Dr. Eidbo began his work. Many of the bites were quite deep. Many of the claw wounds were not only deep but long. As Dr. Eidbo continued to stitch and sew, I, in turn, became quite uncomfortable about it all and began to shake and complain of being cold. My teeth chattered and I turned, as Dr. Eidbo described it, quite white around the gills. The doctor immediately injected some soothing sedative into a vein and I began to relax. He must have used a pain-killer for I don't remember feeling much of anything after that.

As I recall, it took some hundred sutures to sew me back together and probably a pint of antiseptic solution. After completing his work, Dr. Eidbo stepped back and surveyed the result. He nodded his head in justified satisfaction. I too was pleased—that he was through. He then rather rudely jabbed me in the tail with a big syringe filled with penicillin.

"Darn it," I yelled, still dopey from the sedative, "didn't the leopard do enough to me? Why did you have to do that?"

The good doctor roared with laughter. "Shouldn't think that would bother you a bit, Elgin, after all you've been through today." He motioned to Tom to take my arm. "See that he gets home, will you, Tom."

As we stumbled out of his office he managed to be a bit more sympathetic. "I think you'll be okay, Elgin," he said. "Just come back tomorrow so I can remove the bandages and see how you're mending. I'll phone the pharmacy and have them deliver some antibiotics and something to relieve the pain. The leopard did a pretty nice bit of surgery on you, but he didn't hit any vital places, just meat and muscle. The only thing we have to worry about is infection, but I think the antibiotics will take care of that."

Tom drove me home. Needless to say I thanked him all the way for his courage and, particularly, for his big, big feet. If he hadn't acted immediately, the leopard would have been gnawing at my throat and that would have been a most unpleasant ending.

I spent the next six days in bed, flat on my stomach, except for a brief, painful period each day in Dr. Eidbo's clinic. I did not find the process of removing all that adhesive tape and re-dressing my poor back a pleasant experience. Dr. Eidbo managed to keep any infection away, though, and I recovered quite rapidly.

As soon as I was able to return to the zoo, I called the animal dealer and asked him please to drive over and pick up his darn leopard. After my sad explanation of what had happened and why I didn't feel particularly kindly toward his leopard and only wished that he would take it away, he agreed to do so.

"I don't really understand why the cat acted that way, though," he said. "He's always been as good as gold with me. I go into the cage and play with him every day."

I didn't argue the point. Undoubtedly he was telling the truth about the animal's relationship with him. As for me, I didn't wish even to attempt to establish a closer relationship with his animal. We had been quite close enough, I felt.

The animal dealer took his leopard back home, and we later acquired a young cub from another zoo. We affection-trained him, he fell in love with Shannon Leopard, and they had a beautiful litter of little leopards. I'm happy to say, too, that our boy leopard likes me and has never attempted to bite, scratch, or devour me.

Skipper and his trainer, Eddie Anderson, who raised the big chimpanzee in his home and then donated him to the zoo when he was eleven years old.

Performing a social ritual practiced among chimpanzees, Skipper grooms the author and thereby initiates him into the chimp clan.

Chief zoo guide Peggy Riley holds Cleo, an affectionate seven-foot boa constrictor. Cleo was the zoo's first reptile after Elgin convinced the city officials that snakes were not inappropriate in a children's zoo.

Ten members of the zoo staff hold Penny, the zoo's twenty-five-foot reticulated python, while she is being treated for a mouth ailment.

Attired in their Snake Cult costumes depicting literary and historical personalities associated with snakes, such as Eve, Cleopatra, and Medusa, several zoo guides support Toc, a large Burmese python. The women are participating in a Snake Cult pageant, one of the zoo's educational programs for children.

Members of the zoo's Snake Cult perform their version of a rain dance, holding harmless snakes up to the weather gods and pleading for rain. Strangely enough, it began to rain just after this photo was taken, breaking a six-month drought in central Iowa. The performance was repeated several days later and six inches of rain fell in Des Moines, setting a new record.

Zoo guide Sue Hardy poses as Eve with a boa constrictor while Elgin charms Suzie, a Pakistan cobra. Fortunately for Elgin, whose flute playing is not up to Rampal's standard, the cobra is "charmed" not by the music but by the movement of the flute. None of the zoo's reptiles are defanged and Suzie's ability to strike lightning-fast, uncommon among most cobra species, keeps Elgin fully alert at all times.

Elgin picks up Huff Cobra to give the camera a better angle on his hooded profile.

Bill Haast, a world-renowned herpetologist, poses with a huge hooded king cobra at the Miami Serpentarium. Haast has been bitten more than 120 times by poisonous snakes and has developed an immunity over the years to many venoms. The antibodies in his blood serum have saved the lives of twenty persons, including Elgin.

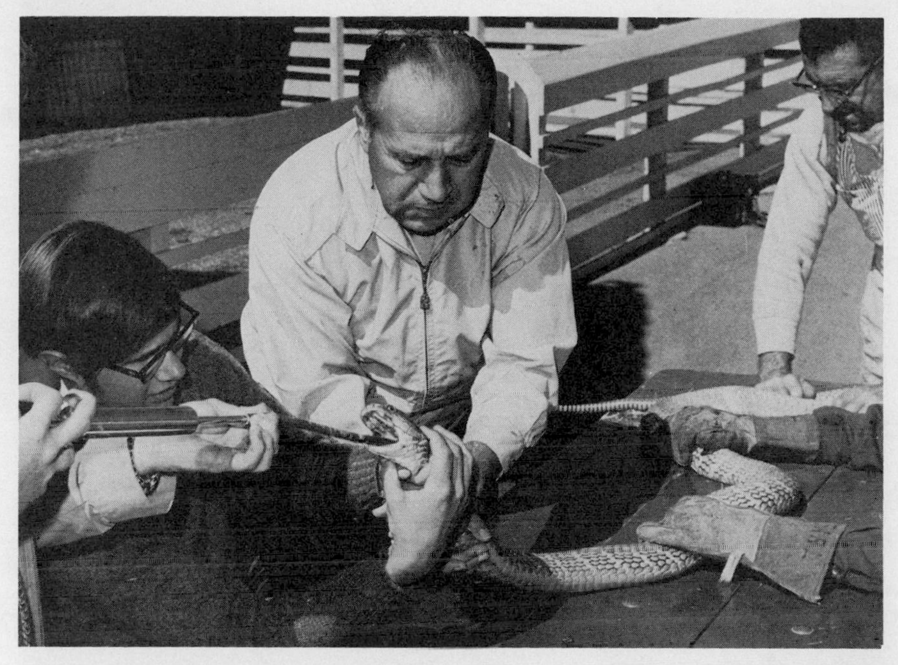

Fifteen feet of very deadly king cobra receives medication for a bad cold—and a meal of baby food as well.

During a reptile lecture, Elgin displays the long, needle-sharp fangs of an eastern diamondback rattlesnake. Rattlesnakes have hinged fangs which fold back against the roof of the snake's mouth when it is closed. Note the drop of venom on the tip of the nearest fang.

Rosie Rattlesnake rests contentedly in Elgin's hands. Elgin has found that even these relatively primitive animals are not instinctively aggressive. Once fear is eliminated, the animal's emotional state reverts to the level usually found in its natural environment. There are no "killer animals." Animals attack only to protect themselves and their young, to eat, and occasionally to protect their territories.

Professor Pedro, a green macaw, campaigns for the United States presidency, representing the zoo's "For the Birds party." When Elgin first became director, the Parks Department would not permit contact between Pedro and zoo visitors because of his vile language. Elgin spent hundreds of hours reforming Pedro's vocabulary by reading to him from Spengler and Korzybski. Having learned such words as "megalopolitan," "transcendental," and "a priori," the Professor adopted Korzybski's non-Aristotelian philosophy as his platform.

Four years later Becky Lion sought the support of the women's liberation faction as the zoo's candidate for United States President. Becky has Elgin "by the throat" but she is merely expressing her affection.

Leonardo Lion, weighing 425 pounds and standing over seven feet tall at the age of three, plays with his human mouse, Director Elgin. Like all of the affection-trained predators at the Des Moines Zoo, Leonardo has been taught not to bite hard and never to use his claws—a result the zoo has been uniquely successful in achieving.

Leonardo attempts to pull Elgin over backwards. Such play is potentially much more dangerous than anything circus trainers attempt, but the zoo's affection-trained big cats—more than fifty of them—have never so much as scratched Elgin in eleven years.

Leonardo has a very special affection for Elgin's wife, Jane.

CHAPTER 8

Becky Lion, Watchcat

One cold, rainy day in early October, lovely Nefer Lion presented us with a litter of cubs. Certainly there was nothing new or different about Nefer bringing cubs into the world since she had done so on four previous occasions. Nor was there anything different about her timing. As usual she had picked the worst possible time of year and the worst possible weather to give birth to the little animals.

Like many zoos, we keep our big cats outside, even during the winter months. As I've mentioned before, lions, tigers, and leopards are extremely adaptable. Despite the fact that they originated in a tropical climate, they can endure and even enjoy very cold weather. It's good for them, too. They have fewer respiratory problems when kept outside rather than in heated buildings. Our lions had a big den to curl up in, of course. The floor of the den contained electric coils to heat the concrete, so the big cats were perfectly comfortable. There was not sufficient warmth, however, in the late fall, winter, and early spring for the baby cubs. And this was always the time of year when Nefer decided to have cubs.

Most probably she knew that we would have to rescue them from the cold and raise them for her, while she, relieved of the responsibility, could continue to sleep all night and play all day.

We knew we must get this latest litter out immediately if we were to save them. As cold and wet as it was, the little animals would lose body heat very rapidly and soon develop pneumonia.

But Nefer had different ideas this time. She loved me, but she was reluctant to come out into the rain, even when I called her. Perhaps, too, a flicker of true motherhood was beginning to burn in her breast and, knowing that I would probably take these cubs away from her as I had previously, she was determined not to come out and visit me and be locked out of the den again.

I called to her time and time again, in lion language and human talk as well. She ignored me. I told her I had lots of fresh meat for her and that she could even play with my hat. She still ignored me. Even Ramses was somewhat upset by my presence. He roamed about the outside area of the cage, being very masculine and protective. Occasionally he snarled at us and told us to go away. He did reconsider every so often, however, and accepted a piece of horsemeat from my hand.

We tried every trick we could think of for two hours, but Nefer refused to cooperate and come out of the den. The rain was pouring down by now and it was easy to see that some of the water was flowing directly into the den area. I was certain the cubs inside must be soaking wet and very, very cold.

In desperation, I sent one of the keepers up to my home (two blocks away from the exhibit area) to pick up my little daughter, Michelle. Michelle was just eight years old at the time, but she had helped raise eleven other lion cubs and she had learned a great deal about lion talk. Particularly important, she had learned to talk lion-cub language in a way no adult could ever master.

When she arrived a few minutes later, we explained to her what was needed. Michelle stood at the far side of the cage where Nefer couldn't possibly see her from the den and made like a lost little lion.

"Owwrr, owwr," she yelped, sounding like a lonesome cub.

She didn't have to repeat the sound. Nefer rushed from the den to rescue what she thought was a stray cub. One of the keepers slid the barrier in place behind her, locking her out of the den.

Unable to find any lost cub and unable to get back inside her den, Nefer graciously accepted the situation and began to eat the meat I offered her. She even permitted me to pat her head and stroke her back to show she forgave me. Meanwhile, two of the keepers opened the outside door of the den and gathered up the three little cubs. They placed them in a box and hurried them out to the warmth of the zoo truck.

The cubs were soaking wet and trembling with cold. We rushed them up to my home and placed them in the zoo's baby incubator. Then we watched and waited. There was nothing we could do for the moment but keep them warm and quiet.

The next morning they were ready to take a little formula from a nursing bottle. Each drank about an ounce—not too much, but we felt encouraged. We fed them again every three hours for the rest of that day and all through the night. They appeared to be holding their own and maybe even gaining a little strength.

The following morning, however, all three of the cubs refused to eat. They were weak and listless and their rapid, shallow breathing indicated an illness we had feared all along. The cubs, like many others before them, had developed a form of chronic pneumonia. From now on it would be difficult to save them.

We first injected doses of commercial feline serum from common house cats into their little bodies to give them some antibody protection and to build up their strength. They regularly received penicillin shots and injections of Ringer's solution to prevent dehydration. Then we began to tube-feed them their baby-lion formula. This is a risky procedure and one I truly dreaded because it's so easy to get the tube in the wrong place and aspirate a lung. Many veterinarians I know share the same misgivings, but the Elgin family did it, fearfully, every three hours for the next three days with each of the three cubs.

Somehow we managed to keep them alive and they improved enough to nurse their formula from the bottle again. We continued to feed them every three hours, morning, noon, and night, and we continued them on medication, for some of their symptoms still persisted.

After two weeks of constant attention, they were out of danger. We reduced their feeding schedule to every six hours, and Jane and I began to enjoy some occasional sleep. Since we had raised four other litters of Nefer's cubs under much the same circumstances, we accepted the fact that they required constant care and treatment, but I confess that I learned fully to appreciate what a mother must go through with a human baby and to be thankful that I hadn't had too much of that responsibility.

By the end of the third week the cubs were too big and active for the incubator. They graduated to our bathtub, where we placed blankets over a foam rubber mattress. The cubs were most comfortable, but it was quite inconvenient for the Elgins. We had no choice, however, having learned previously that we could not place the little lions in either a wooden or cardboard box, regardless of how large it was. They invariably rubbed their little noses raw against the sides of the box. They could not do this in the bathtub because the sides were so smooth.

We began giving them ground meat along with their formula when they were five weeks old. By this time they required more room than the bathtub afforded and they had the run of the house—at least during the daytime. This, as always, was a wonderful experience for us. There's nothing in the world much cuter or more endearing than a fuzzy little lion. The cubs accepted us as part of their lion pride.

They stalked us, leaped at us from behind corners and couches, chased us, caught us, and played with us just as they did with each other. To prevent any feelings of inferiority on their part because of our larger size, we gave each of them a Barbie doll to play with. We weren't anxious to instill a lasting association, but they did enjoy chewing on the small, humanlike figures.

When their feeding schedule was reduced to twice a day, we moved them to a cage in the winter quarters. The Elgins reclaimed the bathtub and the lions began their training. They were taught not to bite hard, never to use their claws, to ride in the rear of my station wagon, to walk on a leash, to enter a transport cage, to visit strange places with confidence, and always to remember that humans were like lions and should be treated nicely.

The process of affection-training big cats and other exotic animals is not an easy one. It consists of 99 percent love, 1 percent discipline, and endless time and patience. It also requires a thorough understanding of the basic psychological dynamics of the animal. Essentially, it's an exercise in preverbal communication between man and animal. The mutual trust and understanding between the two, once achieved, however, create a beautiful and enduring relationship that is well worth the time and effort. Certainly it has helped immensely in our peculiar problems at the Des Moines Zoo. Since their sire, Ramses, was a rare, black-maned lion, we had no difficulty in finding homes for the two boy lions in other zoos. Both left our zoo when they were four months old. But no one, it seemed, wanted Becky, the little girl lion.

This, we felt, was very sad, for Becky was a beautiful and very nice little lion. She loved humans, too. Not only were we unable to find a home for her with other lions, but we, her human friends, often had to neglect her because of the many other necessary tasks we had at the zoo.

We occasionally found time to take her out for walks, but for the most part Becky Lion spent a very lonesome winter. And when spring came, the situation didn't improve much. We were just too busy working with the elephant and making plans for the coming exhibit season. We gave Becky bones to chew on and a small bowling ball to play with, but I'm certain that she was far from happy during that time. She was too much of a people lion.

I felt very bad about this and when she ran to the front of the cage to greet me, as she did each day, I tickled her nose and apologized for neglecting her. Just as soon as I had a little more time, I told her, we'd go for a long walk on the leash and she could roll in the grass and play.

We continued trying to find a home for her at other zoos, but that year, it seemed, no one wanted girl lions, however nice they might be. In the meantime, Becky kept growing, and growing very rapidly, and had virtually outgrown her small cage. Sadly enough, there seemed to be no place in the world where Becky Lion could be truly loved and happy.

Every zoo it seems has its problems with vandalism. Some strange humans seem to find a sense of gratification in molesting caged and helpless animals, teasing them, throwing sticks at them, feeding them inedible objects, tormenting them with sticks, or even killing them with poison or by some other means.

Only a psychiatrist could explain the whys of such destructive behavior, and the prevention problem is even greater. Vandalism occurs in all zoos regardless of size, and while preventing such incidents is difficult in large zoos, it is almost impossible in small ones. Particularly in zoos like ours, which lack the services of a uniformed security guard in the daytime and a night watchman when the zoo is closed.

To say that we were plagued with vandalism is to put it lightly. That spring, before we opened for the season, vandals broke into our McDonald's Barn and sprayed many of the small animals housed there with a fire extinguisher. Some of the animals were killed outright. The chemical affected the nervous system of others and they had to be destroyed. The male rhea and some others lived, but they were obviously affected by the chemical and were ill for a long time.

We were plagued by these unwelcome visitors almost every night. They crept into the zoo area and stoned the animals in their cages, broke windows and doors, tore up the little train, pulled down fences, and did anything else that entered their demented minds. They stole one of my falcons, cut loose an eagle (which we recaptured later because the bird had been trained to return to a falconer's lure), and managed to evade every attempt we made to catch them. Our big difficulty was that we had two different areas to work in—the winter quarters and the exhibit area, which were a half-mile apart—and we had only one man to do the night cleaning and feeding. We needed, as I had insisted over the years, a night security guard.

On one occasion vandals stoned my beautiful golden eagle, Morticia, and broke her wing. She was one of my most prized animals and when she died, despite the best efforts of the veterinarian staff at the State University, I was heartbroken.

Then, just after the season opened, the vandals managed to commit another senseless crime that had me climbing the walls in frustration and anger. They stole a valuable trained hawk that belonged to my friend Dr. David Graham, a professor in the veterinary school at the State University of Iowa.

Dr. Graham had loaned us the bird to complete our predator bird display and I felt personally responsible for the loss. I felt that I had failed miserably in my obligation to him, particularly since there was no way I could replace the hawk.

Once again I approached Mr. VOB with this old problem. "Please," I pleaded, "give us a night watchman. Our one man can't possibly do his work and act as a watchman too. These kooky vandals just wait until he gets into the truck and drives to the winter quarters and then they hit us."

Mr. VOB's answer was the same. He was sorry, but there just weren't funds in the budget for such an expenditure. I could not have a night watchman.

I confess that I got pretty upset about the matter. I was angry all day and so much so that night that I didn't even bother going to bed. I stayed awake thinking bad thoughts and scheming how I might, somehow, embarrass Mr. VOB into making some provision in his budget (immediately or for next year) for a night watchman.

In the early morning hours I found my solution—Becky Lion.

I called all the people in the news media in Des Moines and announced a special news conference at 2:00 P.M. There was something very different about to happen at the zoo, I told them. They all came.

I explained in full detail about the vandalism at the zoo, my inability to obtain a security man, and my tentative plan to use Becky Lion as a watchcat.

"You must be kidding, Elgin," one of the reporters commented. "How would you use a lion to patrol the zoo? Would you just turn her loose inside the fence?"

"No, nothing like that," I replied. "If we use her, I'll walk her around on a leash, or possibly we might chain her out in the area behind the Birthday House where the vandals seem to enter all the time."

"Why don't you just get a watchdog?" another asked. "A lion would probably kill an intruder, wouldn't she? You don't want anything like that to happen, do you?"

"Of course not," I said. "Actually Becky Lion is not nearly as ferocious as a trained watchdog. She's been trained not to bite hard, and never to use her claws. About all she would do is wrestle the intruder down and sit on him if she were to capture one. Then one of the keepers and I would be able to collar him and call the police."

"It sounds pretty drastic to me," said another of the newsmen. "Are you sure the city won't give you a night watchman?"

I shook my head in disgust. "No, I've been trying for five years to obtain one," I explained. "There's never enough money in the budget.

"In addition to the harm that's been done to the animal collection, there's also been a great deal of physical damage to the zoo. Just last winter we had a fire in the winter quarters when lightning struck the building. If Earl Connett, our foreman, hadn't arrived an hour early for work, the whole building and all the animals in it would have burned. That's another reason why I want a night man."

"Well, Elgin," the reporter said, "I'm still a bit skeptical about how a lion would work out, but I confess you seem to have a need for some help out here, even if it is a lion. Can you give us a demonstration of Becky on patrol or something, so my photographer can take some pictures?"

That was a simple matter. I drove to the winter quarters with my son Rob, and we put the collar and leash on Becky Lion, loaded her into the back of my station wagon, and returned to the exhibit area. I led her about the zoo in an endeavor to give the news people an indication of just how efficient she might be.

Becky was only ten months old at the time and she really wasn't very big—just a little over 200 pounds. She took her new job very seriously, however, and acted in a dignified and earnest manner. Certainly she conducted herself as capably as any watchdog I've ever seen. The reporters were all delighted and the photographers took roll after roll of photographs of Becky in action.

The news that we might use a lion at the zoo as a watchcat was carried by TV, newspapers, and radio all over the U.S. and all over the world. The UPI and AP pictures showed our very proud young lioness on her practice patrol of the zoo grounds and stopping for a drink at the big lion-headed drinking fountain. Becky became an international celebrity.

The reaction in the city's Parks Department was much, much different from what I had expected, however. I had hoped to shame them into giving us a night security man, rather than make a poor lion stay up all night and do that work. I was 100 percent wrong.

One member of the Parks Board, I was told later, did suggest that we be given the needed watchman, but the request was never acted upon. Another member of the board expressed the hope that such nightly patrols would not make Becky too tired.

Mr. VOB told me that he didn't think too much of the idea, but that it might serve as a deterrent.

I was dumbfounded. Instead of obtaining a human watchman, I had merely provided them with an acceptable substitute that didn't disturb the budget in the slightest—not even a nickel's worth.

The situation, I decided, was just too hopeless to do anything but laugh about it—and myself.

I was very happy about one thing, though. Becky Lion, our poor, unwanted, and neglected little girl lion, had found her place in the world. She had an important position now and she didn't need to go to another zoo. We had an excuse to keep her. She had found her home.

Mr. VOB had been right in thinking that the news of a lion patrolling the zoo area might serve as a deterrent. It worked perfectly. Our vandalism problems stopped at once and we never had a serious instance again until the fall of 1975. To give an idea of just how effective the watchcat patrol really was, I might use this illustration. Earl Connett was monitoring the police calls on his radio equipment at home one night when he heard the dispatcher request an officer in a squad car to investigate an incident at the zoo. A neighbor across the street from the zoo had called the police and told them two youngsters had just climbed over the fence and were in the zoo area.

"Please check this out, will you?" the dispatcher requested the patrolman.

"Absolutely not," the officer replied, "or not until you get Elgin down to the zoo and he puts that lion watchcat of his away."

The dispatcher called me to do just that so the police officer could enter the zoo. By the time I arrived on the scene, the kids had scampered back over the fence and were far, far away from the zoo area. We never found them.

Actually the police patrolman needn't have worried. We never really used Becky Lion to patrol the zoo. Anybody who knows anything about big cats would realize just how difficult that would have been. Lions and tigers are big, heavy-bodied animals and they don't enjoy long periods of walking or exercise without resting quite often. No intelligent trainer in a circus ever works his big cats for more than thirty minutes, and certainly we never exceed that time when we take the animals out for walks at the zoo. After all, who wants to drag a big, lazy lion or tiger around the zoo all night on a watch patrol. Certainly not me. I'm much too lazy myself to do anything of the sort.

As I stated before, however, the news that we might have a lion patrol at the zoo worked wonderfully. We've designated other lions and tigers for the honorary role of watchcat at the Des Moines Zoo since Becky held that position, and there's no question that this has served to deter vandalism at our little zoo.

Later in the summer we decided to run Becky Lion for the presidency of the United States. She would be the first female to run for the office.

We had nominated the zoo's big green macaw, Professor Pedro, as a candidate for the presidency in 1968. He was the choice of the zoo's For The Birds party, but Mr. Nixon had soundly defeated him. The Professor's non-Aristotelian approach to the nation's problems had been too advanced for the voters to appreciate at that time, and we were afraid they might still lack the ability to understand the great bird's far-out and unique thought systems. The Professor agreed to step down as head of the party and act as running mate for Becky Lion. He would seek the office of the vice-presidency.

Everyone at the zoo felt that Becky would be a perfect symbol of the women's liberation movement and would attract support from all the woman voters in the country. She was certainly beautiful. She was also gentle and very much a lady. Most important, now that she was a big lion she had a certain air of authority and capability that would undoubtedly command respect from every voter.

We prepared our campaign signs. This time we were careful to keep them simple and hard-hitting. We stayed away from the esoteric slogans about non-Aristotelian philosophy that we had used in the Professor's campaign.

"Ms. Becky Lion for President," one sign proclaimed.

"Girl eagles rule the roost," another stated.

"Becky Lion supports planned parenthood—14 brothers and sisters," another banner read, expressing her views on birth control.

Yet another sign urged all the donkeys and elephants to unite and support Becky Lion for President. "Girls are superior," it affirmed.

"Girl elephants lead the herd," another banner boldly declared.

The news people were delighted by the colorful banners and by Becky's platform and forceful personality. Pictures and news stories were again carried all over the world. Becky was again famous.

For the first month our campaign went very well. We conducted a straw poll at the zoo and the results showed Becky well ahead of all the candidates, including the incumbent President Nixon.

Later in the fall, however, it became apparent that President Nixon was gathering strength and gaining over his Democratic opponent. We had hoped that the Democratic party candidate might show more strength and split the vote. Then, we hoped, Becky Lion would be in a favorable position to swing the election.

History records, however, that President Nixon won by a landslide. Our Becky Lion, like Professor Pedro, was defeated.

Becky Lion's campaign for the presidency was the thirtieth time the Des Moines Zoo had received national attention in the newspapers or on TV. All of our promotions had received local and state-wide attention as well. We knew, from the surveys we took at the zoo, and from a later survey taken by the *Des Moines Tribune*, that our public relations program had attracted a great deal of favorable attention among the residents of our city and that we had achieved a tremendous amount of public support. Our surveys indicated, too, that there was a real desire for at least an adequate zoo, if not a big one, and that our visitors wanted a year-round exhibit building and large, natural habitat areas for the animals.

Yet there still seemed to be no possible way of providing these facilities.

Each season our zoo visitors complained more and more about animals in cages, whether the cages were large or small. The handwriting was most clearly on the wall. It was all too evident to me that it was just a matter of time before the big national humane groups would become concerned about the situation and insist that zoos get their animals out of cages and into larger areas. Zoos would then have to shape up or close down. And I had to confess to myself that the humane groups would be, in a sense, quite right.

I became discouraged and despondent. Many wonderful people had taken an interest in our zoo and were helping in every way possible, but raising the large amount of money necessary for expansion was a very discouraging consideration for everyone.

The Parks Department had just constructed a new municipal golf course adjacent to the south side of the zoo. One of the Zoo Association members asked Mr. VOB whether there might be any hope of obtaining funds from the Parks Department or the city to construct a new exhibit building.

"No," Mr. VOB replied, "I doubt that there'll be funds available for that."

The *Wall Street Journal* put it quite succinctly in an article they published on zoos in this country: "Z doesn't stand for Zoo; it stands for ZERO in the budget."

Once again, for the sixth year, we led our animals past the empty, waiting park where we hoped someday to turn our dream of a year-round facility into reality. We continued sadly up the long road back to the winter quarters where the cockroaches and rats awaited our return.

The cage doors were shut and locked behind the animals. The animals and the staff had returned to their winter prison.

My worst fears were realized just a few months later. Mr. Bernard Fensterwald, Counsel to the Committee for Humane Legislation, suggested that the "time had come to phase out zoos." All zoos, big or small, he decreed, were cruel to animals and should be closed out as their collections decreased. Mr. Fensterwald was, of course, merely representing the opinion of many national humane groups in their shortsighted views of animal survival. He and his cohorts actually thought, I believe, that such intellectual concepts as the Endangered Species Act and the animal preserve idea would insure the safety of the animals in Asia, Africa, and South America—despite the incredible human population growth in those areas that mathematically doomed the animals to destruction.

CHAPTER 9

Becky Lion Captures Santa Claus

Just a few days before Christmas 1974 I had a happy thought. The animals should have a Christmas, too, I decided. Santa Claus would visit them and bring them special treats, such as bones and toy balls for the carnivores, oranges and apples for the primates, and special extra amounts of their favorite feed for the hoofed animals.

It also seemed like an excellent way to obtain some publicity in the *Des Moines Register* and *Tribune* newspapers—just to let people know the Des Moines Zoo still existed, even if we were rather shut off from the rest of society during the winter.

If the Christmas party were to receive recognition in the newspaper, though, I knew it would have to have a slightly different angle, a more exciting approach. I gave the matter a lot of thought and, as usual, the answer arrived around 3:00 A.M. I spent the rest of that night embellishing the idea and scheming about how I could get it done.

If, I calculated with much cunning, Santa Claus were to visit the Des Moines Zoo with presents for the animals, it would certainly be very different and most amusing if he were to be captured by our famous watchcat, Becky Lion. Of course he would not be harmed and would immediately be set free to visit the children in the rest of the world, but I thought such a slight misadventure at the zoo might create a chuckle or two in our city.

I arranged for a free-lance photographer to take the pictures, lined up the needed help from the zoo staff, and rented a Santa Claus suit. Both the *Register* and *Tribune* and the local AP office had expressed an interest in the promotion.

Early that afternoon I donned a portion of my costume and had a long visit with Becky Lion. I wanted to make certain that Becky knew it was her friend Bob inside that outfit and not some new, fuzzy type of red and white dinner.

Becky was her usual loving self. She pressed against the inside of the chain link on her cage, asking me to scratch her head and tickle her nose through the openings in the wire. Apparently she was quite aware that it was me inside the suit, though I hadn't as yet put on the red cap and the white whiskers.

She "uurfed" at me in that soft little moaning voice that lions use to express deep friendship and love.

Keepers Tom Sherratt and Mark Morris and I made the final preparations that evening. The photographer arrived and Sherratt drove him down to the exhibit area where Becky was to capture Santa Claus. Mark helped me into my outfit and stuffed my front-middle area with a small pillow I had brought from home.

We were ready now to release Becky from her winter quarters cage, lead her to my station wagon, and drive her down to the exhibit area where the photographer was waiting. Tom Sherratt had returned and I placed him and Mark Morris on top of the long, empty cage next to Becky's. This was a safety precaution we always took when we released a big cat from its cage. The animals are usually pretty frisky when first released and I prefer to establish a little control over them and calm them down before they get close to other humans.

Still without my beard and headpiece, I began to open the door of Becky's cage. Eager to get out, Becky Lion dashed through the door, ran around for a moment, and then ran over to me. She stretched up full-length with her paws on my shoulders and gave me her usual hug of welcome.

Suddenly I felt her abdomen tense and harden. Her claws came out and she began to tear at the costume.

She began to pull me all over the place, clawing and biting at the red flannel material.

"No bite," I screamed at her, time and time again. I pounded on her nose and her head with all my strength, struggling to get free.

I realized the lion wasn't mad. She was simply fascinated, berserk with desire. Something about the costume was driving my beloved lioness goofy. I had turned into an irresistible piece of king-sized catnip.

I knew Becky was just playing with me—the way a cat plays with a mouse or a bag of catnip. She had found a new toy, me, and she was playing with a terrible intentness that was most disturbing.

Mark and Tom were horrified and helpless. There was nothing they could do, no way they could descend from the top of the cage and distract the cat or pull her away from me. They knew, and I knew, that she might resent that and turn on them if they tried to interfere.

"No bite, no bite!" I yelled at her again and again. This was a command she had been taught to obey since she was a tiny cub. She heard and I'm sure she tried to obey, but in her state of intoxication she probably wasn't aware of what she might do to me.

It was obvious that this all-consuming desire of hers was centered on something about the Santa Claus suit and not me. The only difficulty was that I was inside the outfit and, if the costume was getting terribly tattered, certainly too, Elgin was getting terribly worn.

Since the lioness was free there was no way I could get away from her. I was a fat mouse being cuffed about and pulled on by a big three-hundred-pound cat.

We always carried a 22-caliber blank pistol with us when we worked with our bigger cats. Sometimes when trouble occurs, the noise disturbs the animal and makes it change its mind about pressing home a charge. I yelled at Mark Morris to shoot the weapon.

He fired three quick shots. If Becky Lion heard the reports she gave no indication. She was too absorbed with me.

I struggled desperately with her, knowing that if she got me down it might be very difficult to get up again. And I struggled, too, against the fear that was building up inside me. I was in a very uncomfortable position.

Then I managed, somehow, to think of a possible way out.

"Throw me the leash chain," I shouted at Mark, "and hold onto the other end of it."

Mark crawled over to the top of Becky's cage and grabbed the leash. He then tossed one end of it to where I could grab it with my hand.

Getting the snap onto Becky's collar was no problem. Her big mouth was busy chewing on my Santa suit pant-leg and, when I pulled her heavy collar into position and snapped the leash onto it, my action didn't distract her for an instant.

Mark secured the other end of the leash to the top part of the cage with a padlock. At least the cat was secured to one place. Now, I told myself, if I could just escape to another place. Please heaven!

I tried to pull away from her. Becky dug her claws deeper into the material. I pounded on her nose, yelled at her, struggled with all the strength I had. Mark fired the gun again and again, while Tom Sherratt tried to pull the cat away from me with the far end of the leash.

At last something gave—the lower half of the Santa Claus suit. Becky had that entire portion, and I was left standing in my tattered longies.

Mark and Tom descended from the cage top on the other side and helped me to the far end of the building. A quick examination showed that my back was scratched in a few places by Becky's claws and that my legs had a couple of black-and-blue marks. Certainly, we all agreed, I wasn't seriously injured. It was pretty evident, though, that I was scared very nearly to death.

While I sat there shaking like a leaf and almost exhausted, Tom ran across to the main building and returned with my work pants and shirt. I got into them as fast as possible.

In the meantime Becky was still completely engrossed in the remnants of the Santa Claus suit. I don't believe she even knew I was no longer inside it.

We waited a long time. Becky chewed, clawed, and ripped the material into tiny bits. Finally she tired of cuffing at the pieces, throwing them into the air, and mouthing them as though they were the most delectable item in the world. Panting from all her exertion, she flopped down on the floor and looked over at the three of us, tired but happy.

I approached her very, very carefully. As I neared, she sprang up, placed her paws on my shoulders, and "uuurfed" softly at me. She was relaxed and once again her old lovable self. I played with her for a few minutes and, just before I returned her to her cage, she licked my face with her rough red tongue. I'm sure it was her idea of a kiss.

And certainly she owed me that at least. The rented Santa Claus suit she had ruined cost me $75 to replace.

We then drove down to the exhibit area where our photographer was waiting, half stiff with cold. I explained what had happened and he agreed that a Santa Claus attired in work clothing just wouldn't get the message across to newspaper readers. We agreed to try the idea again a few nights later with our smaller lioness, Janie.

CHAPTER 10

The Zoo's TV Stars

Bob Kennedy, host of the top-rated *Kennedy and Company* TV show in Chicago, called one day and asked us to bring some animals on his program. We talked the matter over with the people in the Parks Department and, since Kennedy was paying all of our expenses, of course, they gave the necessary approval for the journey.

I loaded up the zoo's tame rattlesnake, one of our very untame cobras, and Jackie Jaguar. This curious collection of man and beasts flew into Chicago for our first big-time TV appearance.

Kennedy had a real rapport with animals and made us feel very much at home. Unlike some talk-show hosts, Bob wanted to get into the act when the animals were being handled or performing. His enthusiasm gave me one of the most frightening moments I've ever experienced.

I was holding our tame rattlesnake, Rosie, in my bare hands when Bob suddenly decided he would like to have the reptile rattle for the viewing audience.

Reaching over, he seized the snake by the tail and began to shake that part of her anatomy with great vigor. Rosie was too surprised to do anything but turn and look at him.

Me? I almost died of sheer fright!

Jackie Jaguar and Huff Cobra both performed well, too, and Bob invited us to return. He understood our problems at the Des Moines Zoo and sincerely sympathized with our efforts to secure support for it and expand it into a modern zoo.

We made a number of appearances on the *Kennedy and Company* show, until we learned that Bob had suddenly died of cancer in the fall of 1974. He was a remarkable personality and our zoo lost a true friend.

Jackie Jaguar provided an additional bit of excitement on our return home from that first visit to Chicago. The plane had just landed at the Des Moines airport when one of the flight attendants suddenly grabbed my arm and drew me aside.

"Something bad has happened," he whispered. "Your big cat has escaped from its cage and is now loose in the hold of the aircraft."

His statement frightened me for a moment. "Has she hurt anyone?" I asked, holding my breath.

"Oh, no," he replied, "she's just roaming around inside as far as we can tell."

He hurried me down to the side of the big jet where a large group of pilots, stewardesses, and ground personnel were standing in something like stunned silence. A few of their faces were decidedly pale.

"One of the ground crew opened the door to take out the luggage," a pilot told me, "and he found himself face to face with your animal."

The crewman spoke up. "No," he said, "she didn't snarl at me or try to attack me, but I sure didn't waste much time getting the door closed."

We talked the problem over for a minute or two and decided that the best solution would be for them to put me on the rubber hoist and lift me up into the aircraft through another door at the rear of the plane.

Once inside, I could see Jackie lying down at the far end and I started toward her. She was panting, either from fear or exertion, so I approached her slowly, talking to her all the while.

Suddenly she got to her feet and started toward me. I immediately sat down on some nearby luggage and waited, just to let her know that I wasn't about to press her into a bad position. Any jaguar, even Jackie, can be difficult in a strange place, and I wasn't anxious to add to any possible problem.

Jackie, however, was delighted to see me. She felt, I'm sure, that I was her one true friend in the midst of this weird, funny-shaped world with its multitude of strange boxes. She rubbed up against me, pressed her beautiful head against mine, and "neeeaad" in the way jaguars greet their friends.

I noticed then that her front paws were badly lacerated and bleeding. A glance at her cage told me why. Somehow, when the zoo staff had lined the inside of her cage with metal flashing, they had failed to nail down one little spot on the floor. Jackie had found this tiny flaw and clawed the metal loose from the front, then chewed her way through the wood until she had a hole big enough to escape through.

I fastened the lead chain to her collar and gave her a quick hug around the neck. Then I yelled for the crewman outside to open the door. Once this was done, they gave Jackie and me a ride down on the big rubber belt they use to lower the luggage from the plane to the ground. Jackie apparently enjoyed the experience, in spite of her injured paws.

The sight of a big jaguar and a man suddenly appearing from the hold of an airplane on a luggage belt must have astonished many people at the Des Moines airport that day. Certainly the air crew and attendants were very quiet as we descended.

Despite her bleeding paws, Jackie walked the entire half-mile distance across the ramp and into the terminal building. I led her to a cool, quiet corner of the room. She flopped down and patiently awaited the arrival of the station wagon.

It arrived a few minutes later. We opened the rear compartment, Jackie jumped inside, and very soon she was back in her cage, licking her torn paws. Happily, they healed within a short time.

I suppose the moral of this story is that if jaguars, through human error, are to escape from cages in an airplane, they really should be affection-trained jaguars and not mean, nasty jaguars that like to eat humans for dinner.

CHAPTER 11

The Mad Adventures of Brucie Tiger

Another zoo wished to purchase famous Becky Lion watchcat so she could become a mother. Because she was such a beautiful lioness we felt it would be a shame if she were not allowed to have little cubs. She couldn't do that at our zoo, since Nefer Lioness was producing babies at least once or twice a year. We had, if anything, an overabundance of baby lions. We sadly sold her to a very fine zoo; we were sure she would be happy there, but the Elgin family shed many tears the day our wonderful lion friend departed.

Brucie Tiger, who was destined to become as famous as Becky Lion in the role of the zoo's honorary watchcat, was five months old when he arrived at the zoo in September of 1972. I've never seen a more affectionate animal. When we first opened his shipping crate inside the winter quarters cage, Brucie bounced out, voicing his "pfffffff" greeting, which means love and friendship in tiger talk. He raced over to us, raised his beautiful little head, and actually kissed my friend Martin Laugk on the nose. We were delighted, of course. We played with him for an hour or so, fed him a special meat treat, and gave him water for the night.

Then, as we left, I made a terrible mistake. I removed his traveling crate from the cage because it was slightly dirty. Our new tiger then had no place to hide, no feeling of security in his new home. To make matters worse, that inveterate hater of all cats, Skipper Chimpanzee, was caged just across the aisle from little Brucie.

I can easily believe that Skipper spent most of the night throwing pieces of food and feed pans at the young tiger. There were scraps of food and three pans in front of the tiger's cage when I entered the building next morning.

Brucie Tiger was a very different animal from the friendly one we had left the night before. Now he behaved as though he were completely psychotic. Terror-stricken, he raced madly about his cage, certain that his new world was filled with enemies.

As I opened the door of his cage to go in and comfort him, he suddenly darted between my legs and raced down the aisle between the cages. The huge chimp grabbed at him as he raced by, but fortunately missed.

I ran after the tiger, desperately afraid that Skipper would manage to catch him if he again went too close to the cage. Then the chimp would, I knew, either maim or kill the little tiger before I could do anything.

Chasing an animal in such a manner is about the worst possible thing one can do to it psychologically. The proper approach is just to sit down, make yourself small, and let the animal come to you. When tall humans descend by crouching, or even lying down, to the cat's own eye level, he feels reassured, assumes that you want to play, and returns with a happy smile on his face. No big cat can resist this invitation to romp and play, usually, no matter what may be bugging him.

I had no choice, however, but to chase and catch the tiger before he got close to Skipper's cage again. Finally I cornered him, picked him up in my arms, and carried him back to his cage. He gave me a good bite to express his feelings on the matter.

It was obvious that he would never be happy in the same room with Skipper Chimpanzee, so I took him over to an empty cage in the main winter-quarters building. Here, at least, he had only a small macaque and little Daisy Gibbon to look at across the aisle and I was certain their activities wouldn't bother him in the least.

This was the beginning of five of the most difficult months I've ever spent at the zoo. I was determined to regain the tiger's confidence and trust, to make friends with him again. I owed him that much, at least, for the stupid error I had made in removing his shipping box the night he had arrived.

For the next five months I spent four or five hours a day in Brucie's cage without missing a day. I brought bones, wooden balls, nice leather gloves to chew on, choice pieces of meat, everything I could think of to lure him out of that far corner of the cage and somehow make him play again. Brucie crouched in his corner and snarled and spit at me. I sat in my corner patiently, trying somehow to convey my friendly intentions. Every time I attempted to move an inch closer to him, I was rebuffed with a threatening snarl and a show of teeth.

After three weeks I was able to play "snake" with him. It was the first encouraging sign. I used a long strip of heavy leather, wriggling it along the floor of the cage, closer each time to where he was crouched. One day he began to play with it, pawing at it first, then taking it in his teeth and pulling on his end in a tiger game of tug-of-war.

It took three months before he trusted me enough to chew on my boots. The boots were heavy leather, but Brucie by this time was eight months old and a big tiger indeed. I tried my best to keep my toes curled and tucked under, away from the tip of the boots, but Brucie's sharp teeth often found part of a toe or foot and I suffered a very painful pressure bruise.

Two weeks later Brucie was pulling on the fingers of my heavy gloves. Needless to say my fingers were not in the finger parts of the gloves. I protected them by curling them in my palms where Brucie couldn't bite them. The tiger enjoyed this new game immensely but it was hard on the gloves.

However, at no time was I able to touch him. While I was somewhat encouraged by the slight progress I had made, I truly doubted that I could reach him deeply enough ever to affection-train him. After all, my tiger-love project had gone on for five months, and I still found a sullen, snarling tiger confronting me every time I made a move in his direction.

After giving the problem some thought, I decided on a new approach. Perhaps, I told myself, absence might make the tiger heart grow fonder. It was, I confess, not an entirely objective decision. Frankly, I was tired of Brucie's sneers, and my tail was quite sore from sitting on the hard concrete so long each day.

For two days I stayed completely away from his cage. Let him hurt, I told myself. My pride was injured because he had rejected my offer of friendship.

The third day I just happened to pass his cage with a pan of meat. Brucie Tiger literally leaped to the front and condescended to take a piece of meat from this lowly human servant. He then retreated to the back of his cage, turned, and lay down. He began to lick his paws without thinking that maybe he had forgotten something. He had forgotten to snarl.

I used the same approach the next afternoon, ever so casually, of course. Mr. Tiger rushed toward me, made with his let's-be-friends "pfffff" and rubbed against the bars. I had won. He wanted me to come inside and play. Reaching through the bars, I petted his back, his shoulders, and his wonderful big head. Brucie loved it. We were friends again. Apparently even tigers cannot stand loneliness.

When we moved to the exhibit area for the beginning of the summer season, Brucie walked the entire distance on a leash, with me holding the other end. It was one of the proudest, happiest days of my life.

Brucie was appointed honorary apprentice watchcat just after the season opened. He followed Janie Lion around with me, while my son Joel led Janie on her leash. He soon learned his duty rounds and to act in a dignified and serious manner while practicing his security work.

Whenever we had the time and the necessary help, Brucie demonstrated just how well behaved and dependable an affection-trained animal can be. We placed him on a heavy, hundred-foot-long chain that was attached to a stake in the center of our big exercise area. Brucie would then retreat, crouched and slinky, to the far reach of the chain while his human decoy (me!) took a stand directly across from him.

This meant that Brucie Tiger had two hundred feet of free running space before he reached me at the other extremity. Once Brucie was in position, crouched and eager, I gave the signal that I was ready by lowering my body position for just an instant.

The tiger started his charge, then built up his speed over the distance, and hit his target like a cannonball.

The only way a human can survive such play is to change the direction of attack at the last moment. If I had permitted Brucie to hit me directly on the legs or any part of my body, he would have smashed me flat from sheer weight and momentum.

I prevented this by raising my forearm just as he was about to reach me. This offered him a definite target and changed his momentum from the horizontal to a more vertical attack. I was then able to receive most of the impact in a more flexible way and direct him up to a standing position when he took my forearm in his mouth.

Occasionally, even this technique failed when Brucie toppled forward from his full height of more than seven feet. Then old Bob just collapsed like a flattened rug beneath the big bruiser. This always delighted Brucie Tiger. He swarmed all over me, pulling at my pant-legs, tugging at my jacket, and slapping my balding head with his paws.

While I was never in the slightest danger, I did find it difficult to extricate myself from beneath his four-hundred-pound body and struggle to my feet again. He embarrassed me greatly one evening in just this way when NBC national news sent a team of cameramen out from Chicago to photograph our tiger watchcat in a practice demonstration.

Brucie pulled me off my feet when we first started our patrol for the TV crew. While I was down, he completely forgot that he was supposed to act like a dignified police patrolman. He ran back to me, began to play, and refused to permit me to get to my feet. For two long minutes the next night millions of people who watched NBC's national newscast must have laughed themselves silly at the sight of an old man desperately trying to regain his dignity with a big tiger sitting on him. I saw the newscast and felt I behaved much like a frightened, flattened, twenty-legged crab trying to regain some semblance of motion.

This bit of action did demonstrate, though, how Brucie might possibly deal with any unwanted vandal who dared to intrude into the zoo area. And the rest of the performance went well enough as soon as I regained my composure, so the newscast wasn't a total laugh at least.

We were embarrassed even more, however, by a picture that a news service took of a portion of our watchcat routine. The picture showed balding Bob sitting by the closed gate that blocked the entrance to the service road. Bruce Tiger was lying down just in front of me. Above us, hanging from the chain-link gate, a big sign read:

TRESSPASSERS WILL BE EATEN!

The zoo crew had placed it there just the day before, after one of them had seen such a sign displayed in a motion picture. Three college kids had painted the sign and they had agreed that the proper way to spell "trespassers" was "tresspassers." So they painted it just that way. I, truly, had not noticed the sign, but the photographer had felt it might add a touch of humor to the watchcat picture and posed Brucie and me beneath it.

The picture went around the world and we received hundreds of letters suggesting that we learn to spell, or at least permit an English teacher to intrude long enough to correct the kooky spelling. Others stated they were happy the tiger couldn't read or he would be too embarrassed to associate with such illiterate friends.

* * *

None of our watchcats, of course, ever had the opportunity to capture an intruder. As I explained before, we never used them to patrol the grounds. Just the fact that the vandals who had been troubling us thought they were there was sufficient deterrent.

Brucie Tiger, however, certainly earned his keep one day in a unique manner.

All zoos have vandalism problems during the day as well as at night. Our zoo certainly needed a uniformed security man during daytime visiting hours. We had received three bomb threats at the zoo. Zoo visitors sometimes attempted to feed our animals everything from rubber bands to handbags and they were sometimes very resentful if we suggested such objects just might kill an animal. One man threatened to burn down the winter quarters, but we talked him out of it.

The greatest challenge we have, though, is to keep some of the visitors from teasing and tormenting the animals. To add to this problem, our guard rails are not far enough away from the caged animals. It would be easy for a tall, long-armed man to thrust his fingers through the chain-link front of the cage and right into a big cat's mouth, if he were foolish enough to do so.

Some people are that foolish, others are just mean. On one occasion, a group of three very big men and two women began making trouble right after they entered the zoo. To make matters worse, the monkeys (as they often do) extended their arms out of the cages toward the spectators. The three men grabbed the monkeys' arms and tried to pull the screaming animals through the chain link.

The monkeys shrieked in pain and indignation. Foreman Earl Connett hurried over to the scene and asked the big bullies, in a friendly way, to please stop before they injured some of the animals.

His pleas were met with laughter and derision. They pushed him out of the way and walked on down to the cat cages. They had no difficulty, big as they were, in reaching across the barrier and pounding on the chain-link fronts of the cages. This, combined with their inane screams and laughter, frightened the smaller cats, who cowered in fear at the rear of their cages.

Once again Earl Connett tried to reason with the bullies. Again he was met with a rude rebuff and told to mind his own business. Angered by their attitude, Earl told them that if they wished to tease the animals he would call the zoo director and have him take one of the big cats out; they could then try to torment the animal, if they wished, at close range. This statement provoked more loud laughter from the three idiots and some profane words.

I arrived at the zoo just a few minutes later and Earl hurried up to the office to tell me he had a problem. One quick look at the three characters down by the cat cages convinced me that he certainly did have one—a problem that I didn't feel our few little members of the zoo crew could solve in any physical way. And a call to the police would not help because it would take too long for them to arrive. I decided to try Brucie Tiger.

Earl took the heavy chain leash off its hook and we hurried down to the tiger's cage. Brucie was about eighteen months old at the time and he was a big tiger for his age. He tipped the scales at four hundred pounds or more.

The group of kooks gathered around the cage as I entered.

"Whatcha doin', man?" one asked, still laughing as though I were the funniest sight in the world.

"I'm going to take this cat out so you can really tease him, man," I replied. Brucie jumped up to greet me as I entered the door. His huge paws rested on my shoulders, his head towered a full foot above mine.

"Bet that old cat ain't got no claws nor no teeth," one of the dudes shrieked at another.

I looked at him for a moment, then pushed one of Brucie's paws toward him. The big claws were most evident. Reaching upward I pulled the tiger's lip up, revealing the long, lethal canine teeth.

Clipping the lead chain to his collar I motioned for Earl to open the door. The three idiots backed off a few feet.

"Let's get some popcorn, man," one suggested to another.

"Yeah, man, let's do that," the other agreed.

Brucie leaped through the open door dragging me behind him and the five walked quickly toward the popcorn stand at the far end of the zoo.

Brucie padded after them, head low, with a menacing look in his eyes. I struggled to hold him back and to prevent him from getting too near. The five troublemakers retreated the whole length of the zoo area, past the popcorn stand, and right out the zoo entrance. They even neglected to ask for a ticket refund. Certainly they had had an experience they would remember for the rest of their lives.

We returned Brucie to his cage and gave him a big bone as a

reward. Then we sat down and laughed ourselves sick over the whole affair. Brucie had followed them only out of curiosity, of course. He merely wanted to play, but the would-be tough guys hadn't known that. Brucie probably looked very big and formidable to them as he paced along behind them, straining at the leash.

The incident gave us an idea. Well aware that this kind of abuse happened all too often when we were not around to catch it, we decided to try a preventive measure.

The next day we placed a sign on each big cat cage that read:

**ZOO STAFF WILL TAKE
THIS ANIMAL FROM CAGE
FOR VISITORS WHO WISH
TO TEASE OR TORMENT IT!**

We haven't had any takers to this day, of course.

If Brucie Tiger was the hero of the zoo that day, he certainly disgraced himself terribly just a week later.

Brucie attended, so to speak, a birthday party, and he was not even invited. Indeed, it might be said that he was a very unwelcome guest.

My son, Joel, had chained Brucie to the corner of the Birthday House while I was playing with the big cat. We had decided to leave him out a bit longer to enjoy the sunshine and grass while we rested a while. The chain was some fifty feet long, so Joel and I moved toward the front of the Birthday House and sat down out of the playful tiger's reach.

Brucie, disgusted at our reluctance to play longer, bounced around to the rear of the building. We watched him go quite happily, hoping he might find something there to amuse himself with, other than two tired human beings. Joel and I sat there discussing zoo problems without the slightest idea that Brucie was, at that very moment, creating a very new and different problem for us.

When chaining the tiger to the corner of the building, both of us had neglected to notice that the big shutters on the rear of the structure were propped wide open to ventilate the interior for a birthday party. If Joel and I had been negligent in failing to notice this, Brucie made up for our mistake.

Tigers, like all cats, are very curious animals. Peering inside from just around the corner, he saw the long tables within, nicely set with paper plates, cups, and plastic spoons. I'm certain, too, that the pop in the cups, and the cake in the center of one table must have intrigued him.

Since the children had already arrived at the zoo's entrance for their party, the girl guide had closed the front doors to the room and hurried up to greet them. She had also, I regret to state, left the ice cream out in its cardboard container. This, too, may have interested our nosy tiger.

Our girl guide brought the children down to the Birthday House for their refreshments and opened the door. To her horror she discovered Brucie on top of a table, one big foot in the middle of the cake, ice cream all over his face. Plates, cups, and pop were scattered everywhere. The party was a perfect mess. The girl guide did the only thing she could think of doing: she screamed, long and loud.

Joel and I dashed to the door to see what was troubling her so greatly. Brucie looked up as we entered and if there was a look of bewilderment on his face, there was more than a faint suggestion of guilt as well. Deep inside, Brucie knew he was somewhere he should not have been and doing something he should not have been doing.

Joel yelled at him. "No, Brucie, no!" he shouted in a loud voice. That was all that was necessary. Our guilty tiger immediately leaped from the table, through the open shutter, and out of the room. He raced to the end of his chain, lay down, and began removing the evidence, the ice cream, from his mouth and whiskers with his paws. He then licked his paws very thoroughly. I'm not certain he was that fond of ice cream, but I'm sure he didn't want us to discover any of it on him.

The children thought the whole affair was simply hilarious. We were grateful for their reaction and their understanding, for their party was delayed for quite a time. First the girl guide had to clean up the room, more cake had to be obtained, and more ice cream as well.

Joel and I led Brucie, in deep disgrace, back to his cage. Since he had already tasted the carton of ice cream—and ruined it—we decided to give him the remainder of it. Much to our disgust he merely sniffed at it and ignored it. Joel washed the last remaining traces of it out of the cage with a hose.

And just to show Brucie we still loved him and that there were no hard feelings, we gave him a big bone with much meat on it. He whispered his happy little "pfffff" at us and began gnawing away on his gift with great pleasure.

The girl guide who had to clean up after Brucie's little misadventure was not so happy, however. She was quite certain that a bull in a china shop couldn't begin to compare to a Brucie Tiger at a birthday party for creating sheer chaos.

CHAPTER 12

There Is Only the Pack

Wolves have a terrible reputation. Next to man-eating big cats and poisonous snakes, wolves probably rate at the top of man's list of monsters. Stories are legion of wolves attacking and killing humans in Russia, Europe, and Asia. In fact, James Clarke, in his book *Man Is the Prey*, suggests that wolves in those areas have accounted for more human fatalities than the big cats have inflicted on the human race in its entire history. As an example he cites the sixty human deaths attributed to the man-eating wolf known as the "beast of Gévaudan" in France in 1794. He also relates stories of later fatalities in Russia, where an entire village in Siberia was besieged by wolves in 1927 and many persons were killed. The Soviet Union, he adds, has now embarked upon a wolf extermination program. In a recent year, the Russians state, 10,000 horses and 35,000 head of livestock were killed by wolves, and 168 humans were attacked—eleven of them killed. In that same year 30,000 wolves were killed.

Yet for some reason most authorities agree that there have been very few, if any, authenticated cases of a wolf killing a man in North America, at least since the white man arrived here. This is rather strange since the North American wolf is a very large animal, sometimes weighing as much as 175 pounds, hunts in packs, and certainly, if hungry enough, could easily down a man or even a small group of men. And certainly, too, there must have been many opportunities for wolves to have done just that during the past three hundred years.

Let me hurry to state, though, that I'm a wolf lover. Wolves have fascinated me ever since I was a child of twelve and saw the survivors of the last remaining pack of true lobo wolves in their exhibit at Kane, Pennsylvania.

These magnificent animals once followed the buffalo herds in the West and, of necessity, were unusually large and formidable. To a small boy, as I was then, they looked as large as ponies.

The Des Moines Zoo had two large gray wolves on exhibit when I became director in 1967. They were mature specimens and beyond any possibility of being affection-trained at their age. Since they were obviously unhappy in their small cage and virtually wore a bloody path in the concrete with their endless pacing, we gave them to the Iowa State Conservation Commission.

We next acquired two young Canadian black wolves from another zoo. As they were young animals, I hoped to affection-train them, but they arrived with a severe intestinal disease and we had our hands full just keeping them alive during the next twenty days. They did survive, but by the time they recovered they were too old and had been handled too roughly in the course of their treatment (we had to force their mouths open and pop pills down their throats three times a day) to respond to my attempts to make them feel at home, loved, and happy.

I spent countless hours in their cage during the following winter and succeeded in convincing them that humans were not all bad, but it was obvious they would never be relaxed and content in their small summer exhibit cages. We gave them to an animal dealer who found a good home for them in a zoo that had a large open area for them to live in. I was very happy for them.

My dream still persisted of having some happy wolves at the Des Moines Zoo, and in 1973 we were able to obtain two young black timber wolves from another zoo.

It was obvious, though, when the wolves arrived that these, too, were probably too old for training. They were over three months old and this was the first time they had ever been close to humans, except when they were captured and crated. I have never seen two more frightened and defiant little animals. And when wolves are frightened but have no possibility of running away, they simply become hostile and aggressive.

When wolf cubs are taken from their mother just a few days after birth and raised on a nursing bottle, they grow up somewhat tame and manageable. If the person raising them really knows his business, he or she can continue the raising process, hand-raise the animals, affection-train them, and even obedience-train them. The wolf that receives the additional care, love, and training becomes, we have found, one of the most responsive and delightful animals God has ever created.

However, since our two wolves were well beyond the optimum training age, I realized that teaching them to accept humans as friends would be an almost impossible task. I began to look for another zoo which would take them and give them a decent home, hoping that perhaps next year we might be more fortunate in finding a pair of much younger cubs.

The two wolf cubs had found a friend, though. My fourteen-year-old daughter, Becky, had fallen in love with the little monsters.

"Please, Dad," she begged, looking up with big, pleading eyes, "don't send them away. Give me a chance to work with them."

"Becky," I responded, "that's simply impossible. These wolves are too old for affection-training. They're big enough, almost, to eat you up."

"Please, Dad, please," she insisted again.

"Becky," I told her, "if they were lions or tigers it might be a different matter, but you know as well as I do that every authority agrees it's impossible to train a wolf that's over two months old. It just can't be done."

Becky merely looked at me for a long minute. "Just let me try, please, Dad," she kept insisting. "Just let me try . . . okay?"

I gave in, with the understanding that she was to leave them alone long enough to eat her meals three times a day.

Becky was hardly a beginner in handling and caring for young animals. By this time she had participated in the raising and education of four tigers, three leopards, fifteen lion cubs, two cougars, and many other small animals as well. The majority of these occupied the Elgin bathtub during their early days on earth so she was fully aware of the proper care and training techniques in every detail.

For three months Becky became a wolf. She spent from six to eight hours of every one of those days in the wolf cage, talking to her little friends, feeding them by hand, touching them gently when she was able to, and receiving some nasty bites at first when the frightened animals felt cornered or trapped.

Gradually, as their trust in Becky increased, we were able to leash them to a long, tight wire which ran from their cage to an adjoining fence. This permitted them more exercise space and it also accustomed them gradually to the outside world.

Becky persevered and she accomplished one of the most unique successes in human-animal relationships I've ever heard of or seen. By the time the two cubs were eight months old they accepted Becky completely. She had trained them to walk on a leash, took them for long walks in the wooded area close to the zoo, and paraded them over to the lecture pavilion where we talked at length to the zoo visitors on just how nice wolves could be and also how valuable they were in nature's scheme of things.

The Des Moines Kennel Club had purchased the two animals for us. We felt this was a significant tribute to the canine family, since most, if not all, dogs are descended from the wolf. We were further delighted when they provided funds to fence a large open area where the wolves could live and run about to their heart's content. No one will ever know just how much the zoo staff appreciated their generosity. The wolves were the first predator animals in the Des Moines Zoo to make it from a cage into an open habitat area.

Becky had named the wolves Hansel and Gretel, and by the time they were two years old they were very big, Hansel particularly. The three of them had developed an unbelievable love for each other and they often engaged in a helter-skelter game that frankly frightened me to death. First Gretel would run and Becky would chase after her. It was then Hansel's turn to chase and capture Becky. Gretel would then return and, between the two of them, they made a mess of Becky's clothing and, always, her hair. They loved to nip and pull at her hair, but they never hurt her, and they demonstrated their affection by licking her face and hands with their tongues.

We often costumed Becky in a Little Red Riding Hood outfit and sent her into the wolf glen. Our zoo visitors saw a much different relationship between Miss Red Riding Hood and mean old Mr. Wolf than the fairy tale had taught them. The children were delighted with our new version of the story, and we felt we made a very strong point in favor of the wolf family. We demonstrated that wolves are not monsters and, as we explained to the children, that love, patience, and understanding can accomplish miracles even with such supposed monsters as wolves. We also stressed that they should be kind and patient with their pets at home, and that love and understanding could work wonders with their playmates—or even their brothers and sisters—as well. They got the message.

Please be assured, however, that no one but Becky enters the wolf glen now that the animals are mature. They have definitely become one-girl animals. Hansel has stressed this point with bared teeth and most earnest growls on the few occasions when other members of the zoo staff have dared to intrude on their sacred ground.

If the area needs cleaning, or if their water trough must be righted again, it is Becky who does it. This pleases her enormously, for she loves her wolves just as dearly as they love her.

* * *

After Hansel and Gretel were two years old, they became the proud parents of five little wolves. Gretel stayed with the cubs in their big den, of course, while Hansel patrolled the area outside. He took his work of being the protective father very seriously. So seriously that I refused to permit even Becky to go into the glen area with him, much to her disappointment. She could approach and enter the den when the wolves' entrance was closed, however, by entering from the outside door.

Gretel was always delighted to see her and permitted Becky to handle and play with the five cubs while she watched admiringly. Hansel, however, did not like this and he evidenced his displeasure by chewing away at the door in a determined effort to get into the den and rescue his family.

I've read Farley Mowat's wonderful book, *Never Cry Wolf*, in which he describes his adventures with a family of wolves. In one of the most exciting incidents, he descended into a wolves' den, head first, and found himself face to face with mother wolf and cubs. Mrs. Wolf cringed and drew away. Mr. Mowat made a hasty exit and was none the worse for his adventure.

It has been our observation that it was just as well for Farley Mowat that the father wolf was not present at the time. Mr. Mowat might not have lived to write more of his delightful books, and that would be a great loss to us all.

After the cubs were two weeks old, we began to take them, one by one, away from their mother so we could affection-train them. We did this gradually so as not to upset Gretel, and we did it only because we had no other choice. Our zoo's budget certainly would not support five more wolves, and there was no way we could ever handle or capture them when they were older, to ship them to another zoo. It would have been suicide to have tried to take them from the glen area when they were two or three months old by just walking in and attempting to capture them by hand. Hansel would have chewed us into small pieces.

Nor was I about to attempt the capture of seven frightened, running wolves with the zoo's tranquilizer gun. In an area the size of the wolf's glen, it would have been an impossible task and we would have risked killing some of the animals in the process.

So we really had no alternative. We had to take them from their mother and affection-train them for their own welfare. We had taken two of the little animals from Gretel when the wolves made a countermove. Hansel dug a big, deep hole in the ground in the middle of the wolf glen and moved the remaining members of his family into a new home. We were at a loss to solve this dilemma for a week or more. Then it began to rain. Hansel and Gretel moved the cubs back into the den because it was dry there. We took two more of the little animals home to our bathtub. Hansel and Gretel were left with one.

Hansel, I'm sure, was determined to save the last cub, even if it took desperate doing on his part. One night he discovered a weak spot beneath the fence where it joined the corner of the Birthday House. He must have been an engineering genius to have found this flaw in our security setup, but find it he did. To further complicate his job, he had to dig in one precise degree of direction to avoid the ground wire we had placed there to prevent the wolves from digging under the fence. He did this, too.

Hansel dug under the fence and escaped. I'm certain he was looking for that one safe place we could never find, to hide the remainder of his family. He searched all over the zoo area and through the park north of the zoo as well. Some of the friendly persons who live across the street from the zoo saw him running about and they called my home. I was away that evening and both my elder boys were absent, too. Only my two daughters, Becky and Michelle, were at home when the telephone rang.

While Michelle tried to reach us by phone where we were visiting twenty miles away, Becky ran over the golf course that separates our home from the zoo and entered the zoo grounds. She had no trouble finding Hansel. Probably the big wolf had smelled her a distance away and was making his way toward her. They met about a hundred yards from the wolf glen. And they met as friends! Hansel was delighted to see her and expressed his joy by frantic tail-wagging and many wet kisses on Becky's face with his tongue.

It was a simple matter, Becky said later, to get him back into the wolf glen. He followed her gladly the entire distance without even a leash. He walked along beside her and when she opened the door to the den, he licked her face once more and leaped into the den where Gretel and the cub were waiting.

My wife, Jane, and I returned home hours after all the action had taken place. Both of us were frightened when we thought of what might have happened if the wolf had managed to make his way into the neighboring residential area and what could have happened to our daughter, Becky. Frankly, I couldn't believe the story she and Michelle told us. It wasn't that I doubted Becky's word—and certainly not her courage. I just couldn't believe that Hansel hadn't attacked her, injured her, or possibly even killed her.

I had been down to the wolf glen that very afternoon to check on the animals and had witnessed just how frightening an angry wolf can be. As I approached the fence, Hansel leaped at me with every intention of tearing me to pieces. The fence stopped him, of course, and he vented his frustration with snarls and bone-chilling snaps of his long white fangs.

What Becky had accomplished was almost a miracle as far as I was concerned.

Only one problem remained. As Becky pointed out, the hole under the fence was still very much present, and there was nothing to prevent Hansel from escaping again and possibly digging his way under the chain link that bordered the zoo. If he did this, he might terrorize the entire south side of Des Moines before we were able to capture him again. Since it was now 1:00 A.M., I decided to drive down, inspect the fence, and place a rock or two over the hole. This, I hoped, would keep him inside until morning when the zoo crew arrived for duty and we could permanently repair the place.

I drove down to the zoo and up to the den area. Just as I turned the corner beside the Birthday House, my headlights flashed on Hansel as he crawled through the hole and out into the open again. I stopped the car, opened the door, and stepped out. Hansel was about twenty feet away. He saw me, bared his fangs, and snarled. I jumped back into the car as fast as I could. I needed help, I decided. Hansel stood his ground, watching me, waiting, and snarling. I admitted to myself that I was a coward. Picking up the little CB microphone I carried in my car, I turned to the emergency channel.

"This is KWL–0198, the zoo keeper," I yelled, "I need some help here at the Des Moines Zoo. Whoever is monitoring this channel please call the police dispatcher and have him get a patrol car out here as soon as possible."

A voice answered me. "Don't yell so loud," he said. "You're not coming through. What's your problem?"

I calmed down somewhat and modulated better. "This is Elgin at the Des Moines Zoo," I explained. "I've got a big wolf out of his exhibit area and I need a policeman. Please call the dispatcher and have him send a patrol car out. He can enter though the exit gate. It's open."

"Okay, Elgin, I got you this time. I'll give the police a PBX and have someone out there in no time."

Another thought occurred to me. "Please have the dispatcher call my home and tell my son, Joel, and my daughter, Becky, to drive down at once."

"Ten-four," the monitor replied. "I'll give him the message. Good luck."

"Thanks much," I said. "I don't want this wolf running all over the south side. KWL–0198, I'm clear and by the side."

I waited in the car. Hansel waited outside the fence, daring me to get out and come closer. The policeman arrived a minute later and, as he drove up, Hansel retreated through the hole under the fence. The officer and I stepped out of our vehicles, talked for a moment, then approached the fence. The hole was big and deep. That was all too apparent. It would take more than a few stones to close it and keep Hansel Wolf safe inside for the remainder of the night.

Hansel crept close to us, almost invisible in the darkness, and snarled. The policeman and I both retreated to the cars.

"He's a rough customer," I said, more than a little shaken.

"Sounds plenty rough to me," the officer replied. "Look here, Elgin, I know you've got a problem here, but I don't see how I can help you much. As long as he stays inside that fence everything will be okay."

"I don't think he will stay inside the fence for long," I said. "And if he does come out maybe you can fire your gun in the air and maybe he'll go back."

"Don't think that'll do much good," he said. "I think he'll stay inside okay, so I'm going to take off on another call." It was obvious that the man had a very deep respect for Hansel and, I suspect, more than a little question in his own mind as to whether he could even shoot the big wolf in the darkness if Hansel did decide to attack. He walked over to his patrol car, got inside, and, much to my dismay, drove off.

I returned to my vehicle and sat inside, waiting and hoping that Becky and Joel would arrive before Hansel decided to come outside again.

They did arrive just a few minutes later, thank heaven, and Hansel was still on his side of the fence. With Becky present, the wolf relaxed and lay down beside the fence where she could talk to him and scratch his head. Joel and I examined the hole more closely and agreed there was simply nothing we could do that would be adequate, until help arrived in the morning.

The three of us stood there all night long, until seven o'clock in the morning. It was an eerie night. I grew to respect wolves very, very much during that time. Hansel had an uncanny ability to approach us in the darkness without our being able to detect his presence. He was as silent and as invisible in the dim light as a ghost. Suddenly he was there; just as suddenly he was gone. Occasionally we saw his eyes shining in the dark, like yellow sparks of fire. Sometimes he snarled as he approached. Then Becky would talk to him and he'd quiet down and resume the endless patrol of his domain.

It was a long, long night and we were delighted to see the morning crew when they arrived at 7:00 A.M. While Becky lured Hansel to the far side of the exhibit with bits of meat, we placed a length of chain link down to cover the hole from the inside, tied it to the fence, and threw dirt over it. This time we made certain that the ground wire covered all possible angles of escape.

We kept two of the five cubs and affection-trained them. The other three went to other zoos. Becky named the black female cub Akela, and the gray male Gray Boy. In many respects they're even more unusual than their parents. Both were much larger than their parents when they were six months old. They're frequent visitors to our home, where they're treated as part of the family and given the run of the house. Becky has even obedience-trained them to a degree and they walk on leash by her side, heel, and sit on command.

Akela and Gray Boy gave me the greatest moment of sheer pleasure I've experienced during the ten yars I've been at the zoo. It happened one cold day in January. Both Gray Boy and Akela were very big and powerful animals at the time. We took them out for daily walks each afternoon. Perhaps I should say "runs" rather than walks, for whenever you take a big wolf out on the leash you may as well resign yourself to the fact that the first thirty minutes belong to him. Wolves just released from confinement love to run. You can control their direction, of course, by verbal command and by the leash, but Mr. Wolf is definitely going to pull you over the ground, uphill and downhill, so fast you'll think you're flying.

They are incredibly strong animals and I truly believe a 125-pound wolf can pull a human along behind him about as fast and easily as a 500-pound lion can . . . at least for a greater time and distance. The lion will always tire and sit down after the first hundred yards, but the wolf can and will run, full out, for miles.

After the first thirty minutes or so, you can begin to work on their obedience training and they will be sufficiently relaxed to stop, sit, heel, and do as you command. That first half-hour belongs to the wolves, however, for unless the animals derive some measure of enjoyment from their training they may become frustrated and simply bite your head off.

On the particular occasion that gave me such deep personal satisfaction, Becky and Joel were giving the wolves their daily exercise. I was about to board an aircraft to fly to Toronto, Canada, for a TV interview on our affection-training and was attired in better clothing than one usually wears to romp and roughhouse with such vigorous critters as timber wolves. While Becky and Joel disappeared in the distance with their frolicking, furry friends, I waited at the den area for their return.

The two wolves and Becky and Joel were gone a very long time and I began to wonder whether something had happened—such as wolves escaping from their trainers and running into the next county. Then, too, I was beginning to get very, very cold since I was hardly dressed for the chilly weather. I waited and waited and shivered and shivered.

Finally I saw the four appear on the crest of a hill about a half-mile away. They were walking along at quite a slow pace while I stood waiting with cold fingers and colder toes. I decided to hurry them up a little.

"Come, Gray Boy, come, come," I yelled as loudly as possible. The foursome stopped for a moment and I shouted a second time.

Then Gray Boy heard and understood. He tore loose from my giant-sized son, Joel, and he came. Head up, in great leaping bounds he ran toward me, dragging his heavy leash chain behind him. That magnificent animal simply flew over the distance. It was sheer poetry, action in effortless flowing motion. Then Becky released Akela and she followed Gray Boy, flying too in a jet-black blur of grace and power.

Mouths open, smiling, panting with excitement, they came closer. I crouched down and they smothered me, leaped all over me, licking my face and hands, wagging their bushy tails and wiggling in that ecstatic way wolves have of greeting their friends.

Becky and Joel arrived minutes later to join the jubilant ceremony. They helped untangle me from leashes and happy wolves, and we led them, dancing with joy, back to their den area. I bent down, they licked my face, and we closed the gate to the den.

"Dad, I think they really love you," Becky said as we walked away.

And what greater satisfaction could any man ask than to be loved by something as magnificent as a wolf!

My wife, Jane, oldest son, Rob, youngest daughter, Shelly, and six-year-old son, Bruce, drove me to the airport and I boarded the plane for Canada—though I was still in some state of disarray from romping with the wolves.

It was a long journey and, although I love flying, I don't believe I even saw the ground once during the entire flight. All I could think of was the spectacle of those great, beautiful animals bounding toward me, huge in their winter coats, closer and closer, filled with the joy of living and running free.

I chuckled to myself in rather smug satisfaction. For if there is one thing that all wolf authorities agree on, it is that wolves can never be released and recalled. It simply can't be done; this they unanimously state. Yet we, in our ignorance, had been doing it hundreds of times, for years.

CHAPTER 13

Janie Lion: A Love Story

My good friend Wolfgang Holzmair is undoubtedly the world's greatest lion tamer. He works with more lions in his Ringling Brothers circus act than any other performer in the world. He puts twenty big cats through a repertoire of tricks that is unique and amazing.

His performance is all the more unusual to anyone who knows much about big cats, for lions are much more difficult to train and work with in an act than tigers. Wolfgang Holzmair does more with lions than most tiger trainers ever accomplish with their animals. This, in itself, says a great deal about Mr. Holzmair.

Some people frown upon such performances because they feel the acts are cruel and present the animals in an undignified manner. Other individuals, such as the distinguished animal psychologist, Heini Hediger, insist that such performances are an answer to the great problem of boredom in captive animals, that they display the animals' beauty and grace in motion, their adaptability, and high intelligence. There are pros and cons on both sides, but I suspect that a good deal of the negative thinking goes back to the days when animal trainers were suspected of using cruel techniques in making their animals perform.

Let me assure you that no good modern trainer uses any cruelty in training his big cats for a performance. The whip he cracks in the ring may make a loud noise but it never touches the cat. It is largely a matter of showmanship and is also used to get the cat's attention. Modern circus lions are treated with tender loving care. Their trainers provide them with the best possible food and veterinarian attention.

As I've stated before, we don't train our cats to do tricks at the Des Moines Zoo, although some of the biggest zoos in the country do give such performances, simply because I'm interested in doing some basic research on their behavior that requires a more natural, relaxed frame of reference.

But I certainly respect Wolfgang Holzmair for being a very brave man and I also respect him for the care and treatment he gives his animals. They are, frankly, in better physical condition than the cats in many zoos, for Holzmair's lions probably get more physical exercise when they appear in two daily performances than most zoo cats get in a week.

Wolfgang is not only a brave man, he is a very strong individual, too. At the conclusion of his act, he drapes a five-hundred-pound male lion over his shoulders and walks away with it. Or at least he did. Just before he and the circus arrived in Des Moines during August of 1974, Wolfgang began having problems.

He began suffering severe pains in the left side of his chest. The pains continued over a period of time, and Wolfgang began to fear he had developed a heart disease. One day, between performances, he consulted a physician.

The doctor gave Wolfgang an electrocardiogram; the test indicated that his heart was in perfect condition. Perhaps, the doctor suggested, he had pinched a nerve between the ribs in his chest.

Had he, the doctor asked, been doing any heavy lifting lately?

After a moment's thought, Wolfgang admitted that he did lift a rather heavy five-hundred-pound lion on his shoulders twice a day.

The doctor, of course, was flabbergasted. Perhaps, he suggested, Mr. Holzmair might acquire a somewhat lighter lion to carry about. This, the doctor stated, might alleviate Wolfgang's symptoms.

Helena, Wolfgang's lovely wife, concurred completely with the doctor's recommendation. Somehow, she insisted, he must find a smaller lion. The big lion, she told her husband, could sit on a pedestal and smile rather than be carried about on her husband's aching back.

Wolfgang found his new lion when the circus came to Des Moines and he visited the zoo. Apprentice watchcat Janie Lion was nearly two years old, a beautiful young lioness with a gentle disposition. Wolfgang watched her intently for a long time as she performed on the leash, then called her to him. Janie ran over eagerly. It was a clear case of love at first sight. Wolfgang was big, strong, and handsome, and he talked to Janie in a low, soft voice. Wolfgang was delighted with Janie, too. She was not only lovely, but she weighed only three hundred pounds.

Janie joined the Greatest Show on Earth and became a circus star. Our zoo lost a fine watchcat and the Elgins lost another wonderful friend, but we were happy for her success. At least in the circus, as Wolfgang said, she would have a chance to become a mother. We had no room for more lions at the Des Moines Zoo at the time.

And of course we still had big Brucie Tiger, first honorary watchcat to protect us from vandals.

CHAPTER 14

Affectionately Yours, the Big Cats

During the latter part of the 1974 season, we again had a very convincing illustration of the value of affection-training our big cats. We were, as usual, very short of help. The evening crew consisted of only two junior keepers, and one of these had to run the little train. The other keeper had the very difficult task of feeding the entire animal collection, keeping the cages clean, and cleaning the winter quarters where some animals were still being kept. He had to hurry. I, in the meantime, tried to supervise the girl guides, the office, and kept an eye open for those few atrocious zoo visitors who insisted on tormenting the animals.

All of our keepers had been intsructed time and time again, never to open a cage door when feeding an animal. They were told and told and told always to push the feed pan beneath the bottom of the chain link on the cages where possible. In a few cages where there was not sufficient room to squeeze the big pans under the chain link, they were instructed to force the feed through the openings in the chain link itself. They were never to open a door.

Pushing the feed through the side of the cage, handful by handful, took a very long time, and on this occasion the junior zoo keeper elected to try a shortcut. He placed some of the meat through the fence at the far end of the cage, and while Jackie Jaguar munched on that, he opened the main door just enough (he thought) to thrust the pan inside. He goofed!

Before he could possibly get the door halfway closed, the jaguar wheeled and charged for the exit. I was standing some thirty feet away, talking to one of the zoo's girl guides and I glanced over just in time to see the boy desperately trying to stuff the jaguar back into the cage.

I was terrified. Jackie was a full-grown jaguar by now and had not been handled for more than a year. I was certain the big cat would tear the keeper to pieces.

I rushed over to add my efforts—in vain. The jaguar just pushed between the two of us and escaped into the area between the barrier and the cage. The keeper, desperately afraid that she might get over the barrier and injure one of the zoo visitors, simply threw himself on the cat and tried to carry her back to the cage.

Fortunately the lad was big and strong, and he did get her turned in the proper direction at least. I then got her attention and she started toward me. I did the only thing I could do. I stepped inside the cage door hoping she would follow me back inside. And this she did, moving as rapidly as only a big cat can.

The keeper slammed the door shut after her. I was alone in the cage with a very big jaguar and I wasn't sure she knew I was her old friend.

She leaped up, her paws resting on my shoulders. I waited, tense with fear, for her claws to come out and her teeth to bite down. Neither happened. Dear, dear Jackie merely licked my face with her tongue and jumped down to rub lovingly against my legs. Jackie was obviously happy to visit with me again.

I asked the keeper to run to the kitchen and get some pieces of horsemeat. While he was gone, I entertained Miss Jaguar. Never once did she forget her manners. The boy returned with the meat and, while the girl guide lured Jackie to the side of the cage with this special treat, the keeper opened the door and I escaped.

I have a special fondness for jaguars. They've been most considerate and patient with some of our stupid mistakes at the Des Moines Zoo. On another occasion a junior keeper entered the rear portion of one of the winter buildings where we kept several big cats. He placed the heavy feed pans down by the door and went over to the jaguar's cage.

Unlatching the safety bars, he unlocked the padlock and took it off. The cage door was held shut by only a small center pin after the other safety devices were removed. The keeper turned and walked back to where the food pans were and the jaguar lunged against the door. The pin gave, of course, and the cat followed him across the room. The keeper turned as he bent down to pick up the feed pan and saw, to his horror, the jaguar sitting, watching him, just six feet away. Evidently she wanted her food right then.

The terrified keeper fell through the door, slammed it shut, and locked it. He then raced to the office where I was working on my weekly reports.

"I've got bad news, Mr. Elgin," he gasped. "I've just let Maggie out of her cage. She's loose in the barn."

No, he assured me, he wasn't injured. The cat had just sat there and looked at him, evidently waiting for her dinner.

We really didn't have too much of a problem. Maggie was a young jaguar at the time, a year old, and about half grown. We had been working with her regularly and I didn't anticipate any problem in getting her back into her cage.

I entered the room a minute later, and the keeper shut the door behind me. Maggie Jaguar was visiting the other cats in the room, sniffing at them in their cages, playing with paws that the other cats pushed out from beneath the cage fronts, and satisfying her crazy cat curiosity. I picked up her pan of food, placed it in her cage, and she walked in after it. It was as simple as that.

If the keeper had made the same mistake with the usual captive jaguar, I'm certain that he would have been badly injured or killed.

Those winter quarters cages and the small center pin and latch that were constructed to secure the cage doors have proved to be our undoing, almost, on a number of occasions. The difficulty is that when the keeper swings the door closed, the pin and latch will hold it in that position but, if the keeper has other things on his mind, he might forget to padlock the latch and to swing the four big additional latches into their proper position.

To err is human and, as Hediger states in his book *Wild Animals in Captivity*, the wild animal is always on the alert and seems to sense the lapse of attention on the part of the keeper. Then the animal makes his move and either attacks or escapes.

Our animals are very alert at the Des Moines Zoo. They've managed to escape from our faulty cages, through keeper error, no less than three times. Thank heavens they've all been affection-trained.

On one such occasion, Earl Connett ran into my office, his usual matter-of-fact composure quite shaken. He was decidedly disturbed. "Bob," he said, with a heavy sigh, "it's happened again. One of the boys failed to secure the cage properly and the black leopard is out."

The young keeper entered the room at that moment. He was almost physically ill from fear. "Sit down on the bench and tell me what happened," I told him.

He sat, took a deep breath, and tried to control his shaking hands. "I shifted the cat into the next cage and lowered the door between the cages," he began. "I then washed out her cage and left. I closed the door behind me and I'm sure the center pin caught in the latch okay.

Next I washed down the area close to the cage and then pulled up the door between the cages so the leopard could get back in his own." He shook his head sadly. "I just forgot to padlock the center pin and pull down the big latches. I went on into the next room, worked a while there, and then started to walk into the back part of the building again. The leopard was out and prowling around the chimp cage."

The poor keeper was a wreck. He buried his face in his hands in complete despair. Actually, he didn't deserve the entire blame for the matter. The center pin and latch system are inadequate. If a big cat simply pushes hard against the cage door, it is certain to open. Padlocking the door and fastening the latches are additional things to remember and, when a keeper is hurrying to get his job done, he's very likely not to remember everything.

I picked up the phone and called my home. My son Joel answered. "Come on over fast, lad," I told him, "and bring the tranquilizer gun and all the equipment. Bagheera, the black leopard, is out of his cage and running all over the rear room of the building."

"Oh, Lord, no," Joel croaked in reply. "That's not so good. I'll be right there."

I called the police and asked for a squad car to be sent out. Somehow I had the idea that, if things came to the very, very worst, the officer might be able to give us some protection with the shotgun the patrolmen carry in their squad cars. He and Joel arrived at the same time.

While Joel and I prepared the tranquilizer dart, the rest of the crew went to see what the leopard was doing. A small window in the door between the two sections enabled them to see Bagheera inside the room. When Joel and I arrived, Earl Connett informed us that the cat was now behind the cages at the far end of the room.

The situation could hardly have been worse. We couldn't even see the cat, let alone shoot it with the tranquilizer gun from where we were.

"Joel and I are going to have to go in," I told the others. "If we get in trouble," I added, speaking to the policeman, "you're about our only hope. If I yell for help, use your shotgun. Actually, I don't think we'll have any trouble whatsoever. The leopard has been affection-trained, but it's been over a year since we've had a chance even to take him out of his cage."

Joel and I entered the room and proceeded very slowly and cautiously down the long aisle between the cages. At the end of the row of cat cages we stopped, peered around the corner, and found

Bagheera crouched between two barrels in the corner. He didn't look too happy about our being there.

There was absolutely no way in the world we could shoot him with the tranquilizer gun. We couldn't shoot through the narrow mesh on the cage and reach him, and, if we stepped out in the open and shot, it would have been a simple matter of suicide. The leopard was only six feet away, and we had learned long ago that the noise of the gun and the impact of the big dart provoked an instant charge when we tranquilized the big cats inside their cages.

My blood pressure was up, very high. I could tell that by the pounding of my heart and the dry feeling in my throat. I looked back at big, rugged Joel, a question on my face. He ran his tongue over his lips, swallowed hard, and shrugged a shoulder.

"Should we take a chance on him?" I asked.

"Suppose we might as well," Joel answered, "but I don't think it's even much of a chance, if he gets nasty about things."

I placed the tranquilizer gun down on the floor next to the cage. It was worse than useless. Next, Joel and I moved out from behind the cage into the center aisle and sat down on the floor. We did this very slowly and very carefully. Bagheera watched our every move with his bright yellow eyes.

We sat there for a long minute, then I began talking to the cat. "Come, come Bagheera," I whispered, trying to keep the quaver out of my voice. It was an old affection-training command we used when we took the cats out for walks on the leash. I hoped he would remember. "It's all right, old fellow," I continued. "Come, come."

And Bagheera came, cautiously, head low. He had about reached me when he suddenly stopped, snarled, and bared his long fangs. I saw instantly what had disturbed him. The keepers and the policeman were standing in the open doorway. Bagheera just didn't care for their presence and was very expressive about his distaste. I, for my part, did not find it pleasant to be inches away from a snarling leopard with nothing to separate us but thin air, so I yelled for them to leave the room and close the door behind them.

The leopard became calm and reasonable again once they were gone. Coming closer, he rubbed his muzzle against my face and pressed his body against my chest. This made me feel infinitely better, of course, but I can't say that I relaxed a great deal.

"Well, Joel," I said, "he seems to still like us. Let's try to crawl him back into his cage."

We proceeded to do just that—to crawl on our hands and knees the entire length of the room, some eighty feet, toward the leopard's cage at the other end. We had no choice, actually. We knew we had to get him past the chimp's cage without the cat becoming unduly interested in the chimp, and the only way we could hold his interest was to keep our body position low, on the same level as his. Our crawling meant "play" to the leopard. Standing up might possibly have been interpreted as a "threat" posture to Bagheera when we passed the chimp's cage.

Foot by foot, we covered the entire distance on hands and knees with the black leopard pacing happily between us. We reached his cage, Joel crawled on in, and the leopard followed him. I, in turn, lured him into the shift cage, lowered the guillotine door between the cages, and Joel walked out—safe and sound.

I'm inclined to be something of an emotional individual during, or after, periods of stress. I grabbed my monster-sized son and hugged him. It was the only way I could adequately express my unbelievable sense of relief and happiness.

We thanked the policeman and he left. Actually I think he was happy to get away from the building, the big cat, and the zoo. Then Earl Connett asked me a disturbing question. "Bob, were you really expecting the policeman to be of any help if the leopard attacked you and Joel?"

I turned and looked at him. He was dead serious. "Well, yes," I replied, "I guess I was, Earl. After all, why not?"

"Don't you realize that those officers carry sawed-off shotguns and that the pattern they make from even thirty or forty feet away would have killed you and Joel at the same time the pellets hit the leopard? The only thing he could have done if the cat got you down was to run up and place the shotgun right on the leopard's head and then pull the trigger." He paused for a moment and raised an eyebrow. "Do you think he would have done anything like getting that close to a killer cat?"

We both looked at each other for a long minute. Each of us knew what the other was thinking. In the future I don't believe I'll ever depend on anyone carrying a sawed-off shotgun to protect me if I get in a tight spot. Any leopard is safer than a cannon like that.

On yet another occasion, Keeper John Hayden slipped on the ice while feeding our big girl tiger, Shelly. His arm went through the bars and into the cage. Shelly seized his wrist and hand in her teeth.

"No bite, Shelly! No bite!" John screamed in a very frightened voice.

Shelly remembered her training and the "no bite" command. She instantly released his arm. John pulled the limb out of the cage. There wasn't a mark on him. The usual captive tiger would have torn his arm to pieces, then pulled Mr. Hayden into the cage, bit by bit. Shelly Tiger had been affection-trained, however, and John Hayden wasn't injured.

There has always been some question in my mind as to how long an affection-trained animal would retain his trust and happy regard for humans without being worked with, at least occasionally. After all, professional circus trainers work their animals daily for months and years on end, but the pressure of raising and training new animals continually didn't permit us to handle and work with our older cats at all. It was apparent that they were always happy to see us when we visited them daily over the years, but I was never certain that they could be fully trusted, particularly if they were to escape from the relative sanctuary of their cages. Yet experience at our zoo clearly indicates that the animals seem never to forget their training, no matter how long it's been since we actually took them out and worked with them. They seem to know we're still their friends and have given every indication that they still trust and love us.

The problems and dangers presented by big cats and other predators who escape from the confines of their captivity seem to be a matter of unbelievable controversy among those who should know something about the subject. Dr. David Taylor, in his book *Zoo Vet*, seems to adopt a very casual attitude. He reflects nonchalantly about the way escaped lions seemingly gather under the shade of a tree and lie happily about, waiting for someone to come along, shoot them with a tranquilizer gun, and carry them back to their cages. He fails to mention, however, whether the individual aiming the tranquilizer gun was standing, gun pointed, out in the open just a few feet from the yawning lions, or whether the man was shooting from the turret of a Sherman tank.

The famous animal trainer Clyde Beatty took escape incidents much more seriously. Beatty would certainly qualify as an authority on such matters, too, for in the days when circus cages were constructed of iron rods in wooden floors, the big cats experienced no great difficulty clawing the rods out of position and running away. Beatty apparently had about as many escapes as he had animals during some periods of his long career. And one gets the impression from reading his books that he had several very close calls while attempting to get the cats back into their cages, and certainly did not enjoy the process in any way.

Pat Derby, in *The Lady and Her Tiger* (written with Peter Beagle), relates many incidents of trained animals escaping from confinement, and it appears as though most of the trainers were not overly concerned about the matter. Of course, as Ms. Derby suggests, most of these animals were altered in one way or another and certainly presented less of a threat to human life and limb than a well-armed cat might do. Then, too, the animals in Ms. Derby's book were probably worked with every day and that also would make a very big difference in their attitude toward anything human.

It is interesting to note that neither Taylor nor Derby ever experienced anything like a serious attack or mauling by a predator animal. Beatty certainly did, several times. The great tiger trainer Charley Baumann was almost killed by one of his feline pupils and, while he relates no great difficulty in persuading several escaped tigers to creep back into waiting cages, he does qualify the matter by suggesting that if a cat is at loose for a period much over one hour, things might rapidly become much more complicated.

I'm inclined to go along with Baumann, particularly after talking at length with Wolfgang Holzmair and many other trainers. I can't just sit and watch with any degree of objectivity a fully armed big cat that is obviously nervous and unhappy about what to do with its newfound freedom. I get up-tight about the matter. Experience and conversations with others lead me to believe that what happens depends very much on the animal, its age, the time of day, how many humans are about and what they might do, and many, many other intangibles.

I confess, here and now, that I'm a great coward and that animals who have escaped from their cages frighten me. I don't like the abundance of intangible possibilities that might upset them and trigger the awesome aggressiveness that could easily result in someone's being very badly injured or killed. Probably, too, I'm still very much influenced by the nasty mauling I received from that big male leopard. I still have a multitude of scars on my body, and the memory of the incident continually reminds me that large predator animals can easily inflict awful damage on the weak and frail human body.

As you can believe, I'm most grateful that the animals at our zoo are nice and affectionate. When some of them have escaped from their cages they've been most considerate. They've remembered their training and retained a happy regard for us humans, even though some of them had not been handled for more than six years. Truly, I find this almost unbelievable, and I feel it speaks very highly for the intelligence and emotional stability of these so-called lesser animals.

Perhaps the biggest advantage to affection-training animals is the relative ease of treating them if they become ill or are injured. A tame animal will accept food containing medication from one's hand without hesitation or suspicion. An untamed animal is likely to be more hesitant and often may not eat the food at all.

When Angela Jaguar became very ill years ago and was unable to eat her regular diet, I was able to enter her cage and feed her human baby food from my hand. It was the only thing she would eat. I fed her daily in this manner for over a month and she recovered.

Mercury Cougar developed an intestinal infection when he was eighteen months old. Our veterinarian prescribed daily injections of antibiotics for a six-day period. Since Mercury was in a large cage, using a long syringe pole was out of the question. We tried it, but Merc was too suspicious of the instrument and easily kept away from it.

Certainly I didn't wish to put him through the traumatic experience of dosing him that often with a medicated dart shot from the tranquilizer gun. Poor Mercury would never have been the same again. Actually, it was a simple matter to enter his cage, attach the leash to his collar, and have one of the keepers hold the leash outside the cage. I figured this would hold Mercury's head, mouth, and teeth away from me while I jabbed him in the flank with the hypo needle. The precaution wasn't really necessary. Mercury took his medicine like a big, brave cat each day and, when we undid the leash from his collar, he purred happily at me. Never once did he snarl or show any resentment.

On another occasion Mercury's left eye became terribly swollen. It was impossible to tell whether he had an infection or, as we suspected, had been stung by a bee. The condition persisted for two days and Mercury was decidedly unhappy. He kept pawing and rubbing his eye until I decided a closer examination was necessary. Rather than sedate him with a tranquilizer gun, I entered his cage, purring as loud as I could to convince Mercury my intentions were friendly. Bending down, I opened his eye and, to my relief, discovered there was no eye infection. Something must have bitten or stung him on the outer lid. Mercury patiently permitted me to daub some ointment over the swollen area, and the next day he was as good as ever.

Janie Lion injured her shoulder against a sharp corner or projection in her cage. The result was a deep puncture wound that developed into a huge abscess. Because Janie had been affection-trained, we were able to open the abscess with a scalpel, clean the area, and give her daily injections of penicillin. Janie seemed to know we were trying to help her and never resented our efforts, although our ministrations must have caused her no little pain at times.

Training our elephant and camel paid off as well. During the terribly hot summer of 1977 we had an epidemic of eye infections at the zoo that were apparently transmitted by flies. Both Bomba Elephant and Kathy Camel succumbed to the disease and we had to squirt an antibiotic solution into their eyes several times daily for weeks. If the animals had not been trained, the treatment would have been impossible and both animals would undoubtedly have gone blind.

Two of our wolves also caught the eye infection. I've often wondered how one could humanely capture, or restrain, any animal as sensitive as a wolf in a squeeze cage and then gently lift the eyelid and administer medication twice a day for two weeks.

We certainly had no problem with our wolves. It was a simple matter to hold the big animals in my lap while Becky administered the medicine. And when it was necessary for the veterinarian to give Akela Wolf a shot of antibiotics to help her along the road to recovery, she took her shot like the big, brave wolf she is—sitting on my lap while Becky held one of her paws.

As Professor Hediger put it so very well: "Tameness is healthy, tameness is expedient." To go a step further, one might also conclude that, on occasion, tameness may even mean the difference between life or death for an animal.

CHAPTER 15

The Happy Christmas

The weather was perfect. The snow had been falling all morning and now covered the zoo area with a soft white blanket. The castle, with its many-colored bricks, glistened like a fairy palace through the snowflakes floating down from the sky. The Old Mill, McDonald's Barn, and the gaily colored roof of the Birthday House were capped with white. The zoo was a lovely sight—a wonderful world of color and beauty and snowflakes—perfectly decorated for our annual Christmas pageant for the children of Des Moines.

Inside the kitchen, warmed by a portable heater, the Elgin family and all the members of the zoo crew were donning their costumes for the big event. We were hurrying as rapidly as possible, for hundreds of children had already arrived and there was still much to be done.

At last the Wise Men were attired in their long robes, bearded and ready. The little drummer boy, Bruce, and his friend Mike were dressed and waiting. Mary and Joseph were in their cloaks for the crèche scene and the attendants were clothed in their oriental costumes.

We walked from the kitchen to the area around the Birthday House where all the animals were waiting. The procession formed in a long line, the gate was opened, the Christmas music filled the air, and the pageant began.

The drummer boy, riding a tiny Sicilian donkey, led the way, followed by his friend Amahl on another donkey. A third donkey, laden with chests of precious gifts, followed close behind.

Next in line, dressed as the first Wise Man, I led Kathy Camel on a long lead rope. Kathy walked softly through the snow, regally even, in time to "We Three Kings of Orient Are."

Second Wise Man Norm Smith came next, carrying Thor, our great African eagle, on a high perch that rose above his head. Thor was resplendent in his plumed hood. The falcon bells on his legs tinkled as Norm carried him along.

Shelly Elgin, cloaked in the attire of a Biblical princess, rode her white pony, accompanied by attendants who walked beside her.

Just behind her, Becky Elgin led her two wolves, Akela and Gray Boy.

And at the rear of the procession, the third Wise Man, Joel, led big Leonardo Lion on his leash. Leonardo was his usual regal self, magnificent and kingly. Some of the visiting children gasped at the size of the huge footprints he had made in the snow as he passed by.

We walked in a big circle through the zoo area and gathered around the base of Noah's Ark where Father Churchman, the priest of nearby Christ the King Church, was waiting. Opening a small book, he read a short chapter and bestowed the blessing of the Catholic Church on the animals.

The procession filed around the castle and entered the elephant's summer paddock. The small building had been brightly decorated with Christmas lights and framed with thick gold, silver, and green garlands.

Mary and Joseph were seated on the hay inside the shelter, bent over the small cradle that held a doll representing the infant Jesus. Our little calf and three small goats were tied close by.

The Wise Men with their attendants and the animals entered the enclosure and knelt down before the crèche. Our narrator read the Christmas story over the PA system, and while the Wise Men presented their gifts to the baby Jesus, she began to explain some of the significance of the pageant.

"The Sicilian donkeys the drummer boy and Amahl are riding are actually the same kind of animals that were ridden in the holy land during the time of Christ," she told our visitors.

"The camel, of course, is the animal that carried the Wise Men from the East, following the Christmas star to Bethlehem, where Jesus was born.

"The three Wise Men were kings of the Orient," the narrator continued, "and undoubtedly used eagles and other birds of prey in hunting and in the sport of falconry. Many of the Renaissance paintings of the nativity scene depict the Wise Men with trained falcons and hawks on their hands.

"Leonardo Lion," she told the spectators, "is present because some of the ancient kings actually used trained lions to hunt with, or to protect themselves, like huge guard dogs, when they went into battle.

"As for the wolves—well, the wolves are part of the pageant because they love to be out in the snow and we didn't have the heart to leave them in their cages while all the other animals participated in the pageant."

After this brief explanation, we all rose to our feet and paraded

Zoo guide Mary Stevens holds a baby ocelot given to the zoo by its owners when they discovered the beautiful little cat could bite very hard and scratch very deep. Elgin strongly cautions against keeping predator animals as house pets, for the welfare of both the owner and the animal.

Young Becky Lion stops for a drink at the lion's head fountain at the zoo.

Keeper Gary Enfield holds his favorite cat, Nefer Lion. Although she is only eight months old, Nefer is an armful—two hundred pounds worth. Gary helped Elgin train Ramses and Nefer, the first animals affection-trained at the zoo.

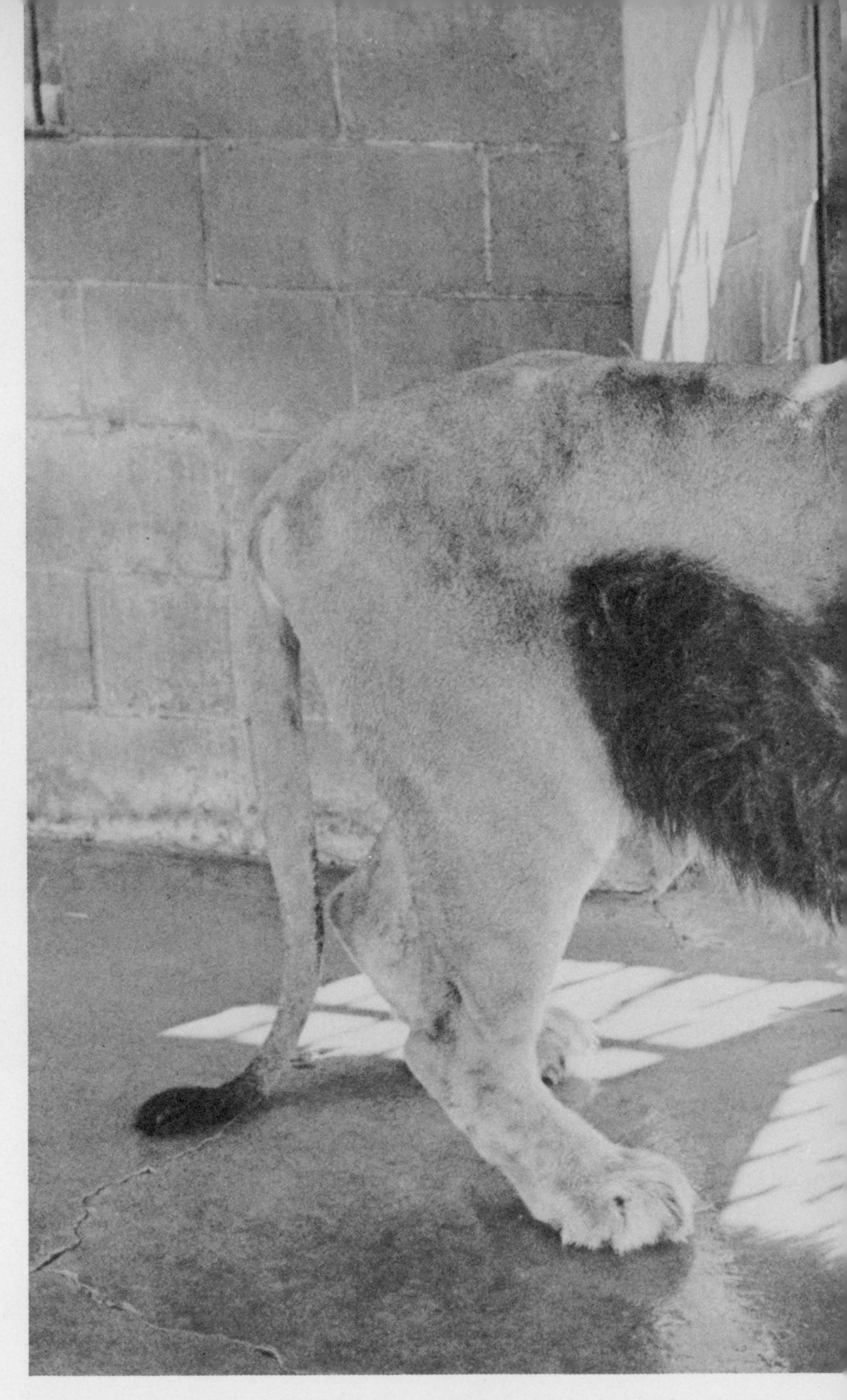

Ramses is the king of the zoo's lion pride and has sired twenty-six cubs over the past nine years. His unusually large black mane, extending over his abdomen and halfway up his flanks, suggests a possible relationship to

the Atlas lion which is now extinct in the wild. Ramses' parents were part of Emperor Haile Selassie's private collection.

Gary Enfield received an affectionate greeting from Shelly Tiger the day she arrived at the zoo. The zoo staff found tigers to be more intelligent and less aggressive than other felines, tending to form a deeper bond with humans.

In a photo that was featured in newspapers across the country, Honorary Watchcat Brucie Tiger poses with Elgin in front of the service gate after making his rounds. The zoo received an avalanche of mail referring to the misspelling of "trespassers" and suggesting that Brucie deserved more intelligent companionship.

Puffing with exertion, Elgin supports Brucie Tiger during an exercise period.

Beautiful Jackie Jaguar pauses for a rest after a walk around the zoo. Elgin has succeeded in affection-training Jackie despite the fact that jaguars are rarely used in circus performances because of their alleged unreliability.

Becky Elgin with three of Nefer's cubs: Luff, Bo, and Top Cat.

Janie Lion surveys her domain as a cub in the Elgin home. Most of the zoo's cubs were raised at the Elgins' because they were born at the wettest, coldest time of the year and the heated lion dens were not warm enough for them. As soon as they could tolerate the weather, however, they were returned to the zoo.

Elgin got an enthusiastic welcome from an older Janie Lion when he spent three days living with the animals to protest statements made by a national humane society that zoos were cruel to animals and should be abolished. The live-in took place just three weeks after Skipper Chimpanzee bit off two of Elgin's fingers. Skipper was donated to the zoo when he was eleven years old and was not affection-trained by Elgin.

Bruce Elgin attempts to hold Mary Lion still for the photographer. Actually, affection-trained cats should never be restrained in this manner. Human trainers should play very much the same role that parent felines do in their natural habitat, displaying warmth and patience when the cubs become overly playful and aggressive.

Elgin leads Bomba, the zoo's Indian elephant, around the paddock while a young zoo visitor enjoys a free ride. The zoo staff trained four Indian elephants and one African elephant to perform this difficult, tiring task.

A young zoo visitor is obviously a bit dubious about riding "tall in the saddle" on Kathy Camel's back.

Zoo guide Kitty Robinson hitches a ride on Barnaby Tortoise by enticing him along with a carrot.

Akela Wolf gives her trainer, Becky Elgin, a wet kiss. Despite their notoriety as animals too wild to tame, Becky has affection-trained twenty-six wolves during the past six years to walk on a leash, ride in the Zoo Mobile, heel, sit, and to respond when called from as far as half a mile away.

Becky is the Alpha wolf in this photo, and Akela's mate, Gray Boy, is momentarily contented with his Beta position in the pack. The struggle for dominance between Gray Boy and Becky has been a fascinating animal behavior study.

Shelly Elgin, attired as Little Red Riding Hood in a scene from the zoo's "Animals and Man" pageant, gets an affectionate kiss from Akela Wolf. The obvious message here is that wolves are not the monsters that many fables have portrayed.

Shelly with a playful litter of wolf cubs.

Peggy Riley with Squeaky, a baby raccoon that has made a home in one of the buildings on the zoo premises.

once again to the area around Noah's Ark. There, one of the zoo crew, clad in long flowing robes and a great white beard, walked out to the prow of our little Noah's Ark and, like the Biblical Noah, released a white dove into the air.

Then suddenly, Santa Claus appeared, high on the top of the castle. The music of "Jingle Bells" rang out over the PA system, and Santa waved and greeted the children standing below. A minute later, when everyone had gathered about, he descended the inside stairway and walked down the ramp into the crowd of eager, excited children. He opened his huge pack filled with candy and gave each child a candy cane and other goodies.

When the candy was gone and the festivities had ended, we put the animals back into their shelters, gave them special treats, then relaxed in the kitchen with hot chocolate and coffee. The pageant had been a tremendous success. The children had thanked us and told us how greatly they had enjoyed it. This made us very happy, of course, but we, too, had truly enjoyed participating. Each of us, I'm sure, had experienced a deeper sense of the wonder, the beauty, and the serenity that Christmas means to many of us.

CHAPTER 16

Our Wolf Pack, Alpha and Omega

Animals are very patient with our bumbling human efforts to understand and help them. We know very little about their real needs and desires. Gray Boy Wolf, for instance, had a habit of pulling his big food pan over to one corner of the den area where he could enjoy eating it by himself. Once this was accomplished, he inevitably turned the pan over and dumped the contents onto the hay bedding in the den. This somehow bothered me, for I didn't feel that he should be eating the little pieces of hay that adhered to the meat. It didn't occur to me that I might be projecting a human concept of proper sanitation that didn't apply to wolves at all.

One day when I was in the den, Gray Boy started to drag the pan of food away to his private corner. I put my foot on the pan. Gray Boy looked up at me curiously—and growled.

He again tugged at the pan. My foot still prevented him from moving it. Gray Boy growled again—this time a deep, earnest type of growl. I hesitated but, determined that I must remain the Alpha wolf in the relationship, I kept my foot on the feed pan.

Gray Boy pulled hard on the pan once more. I bent down to say "No!" to him. Gray Boy reacted instantly. Reaching up, he seized my face in his big mouth. His teeth touched my skin, but ever so gently. I found myself looking deep into the interior of the big wolf's throat. He held me there for an instant, then released me. There wasn't a mark on my face, but Gray Boy had made his point. We were friends, we loved each other, but we were, after all, individuals, and as a wolf he was determined to eat his meal where and how he wished.

I lifted my foot from the pan. Gray Boy tugged it over to his corner, dumped the contents, and ate his dinner with great relish and satisfaction.

I had my moment a few days later, however. Gray Boy had taken his evening ride in the back of my station wagon and, when we returned to the zoo and opened the rear door for him to jump out, he was still engrossed with the big bone he was gnawing on. He continued to chew on it, despite our suggestions that now was the time for him to return to his den, where his dinner and the lovely Akela were both waiting for him.

We called, we commanded, we begged and pleaded. Becky got down on her hands and knees on the ground and made like a playful wolf. Gray Boy ignored us completely.

Now there are certain things that one just cannot take away from a wolf—or from a domestic dog for that matter. One of these is, of course, a meaty and delicious bone. We often had to stop the wolves while walking them and pry their mouths open to retrieve golf balls, plastic containers, and all sorts of things they wished to gobble up. Never once did the wolves object to this, but they were always most possessive about their beloved bones. So please understand that I was not about to reach into the rear of the station wagon and try to take Gray Boy's nice bone away from him. I'm much too big a coward to do anything like that.

It was late and both Becky and I were hungry, but Gray Boy refused to relinquish his bone and jump out of the car. In a last desperate effort to entice him from the station wagon, I uttered my most appealing wolf howl. I intended it to be an open invitation for Gray Boy to join the rest of the pack—Becky and me—on a short journey to the wolf glen. Instead, it had a strange, instant, and dramatic effect on Gray Boy. He immediately dropped his bone, rose to his feet, and stood at attention, ears forward. I howled again and Gray Boy leaped from the vehicle, prostrated himself on the ground, rolled over, and turned tummy up in his most submissive wolf attitude.

I bent down and scratched his tummy. He licked my face and hands. He then jumped to his feet and the three of us ran over to the den, where Akela and dinner were waiting.

After that strange experience I've often wondered whether the howl of a wolf might not mean something more than just an invitation to join the pack for a hunt or a bit of social play. Certain howls probably mean just that, but other overtones or inflections may indicate territorial status and a warning that other wolves should not venture into the howler's territory.

I may have unwittingly made just the proper territorial howl, and Gray Boy may have thought he was directly in the middle of another wolf's territory—mine.

In any event, I confess that it did something for my ego after the way Gray Boy had dominated me in our little disagreement over the food pan a few days before.

Probably, too, I derived particular satisfaction from my triumph because I'm a lycanthrope, a wolf in human form. Or at least my dear friend Jo Sayers, a fellow wolf afficionado, insists that I am. She believes that the peculiar birthmark in the center of my chest indicates that I'm a true and confirmed lycanthrope. I believe I'm quite happy about that.

Every once in a while one of our ducks leaves its happy home in the little pond adjacent to the wolf glen and flies into the wolves' territory. I haven't the slightest idea why they would pick such a place to land, but let me assure you that they do not take off again. The wolves move like lightning, and the duck is soon nothing but a little pile of feathers.

Akela, particularly, has acquired a definite interest in ducks, and this sometimes complicates the daily routine of leading the wolves about the zoo area for their training and exercise.

The ducks roam freely through the entire area, and when Akela spots one she always manages to break away from Becky and chases the duck with all her heart and soul. Despite Becky's yells of protest, Akela has been successful in running down and capturing four ducks. I have no idea why the ducks don't just fly away, since they're perfectly capable of doing so, but for some silly reason they seem to think they can outrun an animal as fast as a wolf. In this they make a great mistake.

Yet the four incidents have not been as tragic as one might imagine. For Becky has always managed to arrive in time to rescue the poor, squawking ducks—after a long, frantic run.

It's really quite simple, at least for Becky. All she has to do is point her finger at Akela and yell, in a very loud voice, "No, Akela, no! Drop that duck this very instant. Drop that duck, Akela."

And Akela drops the duck. The duck runs away, shrieking its head off, but quite unharmed. Akela watches for a moment, then turns to Becky, ashamed and apologetic. She then performs her tummy-up act of submission and, while Becky scratches her, Akela licks Becky's face and hands.

In analyzing Akela's behavior, I'm convinced that it's most unusual for a wild animal to behave in such a manner: to give up a prized possession like a duck without becoming aggressive and hostile.

I once, for instance, owned a tiny fox cub that had been taken from its den before its eyes were open. As an experiment, I gave the little animal a fresh rabbit leg to chew on. The fox simply went crazy over the tidbit and, although the cub weighed only four or five pounds, I was actually afraid to get close to him until he had finished the rabbit.

Instincts are powerful forces in the emotional makeup of many animals. Yet Becky was able to persuade Akela to drop the duck with only a vocal command.

There is no question that Becky dominates her wolves—usually. She does this with love and a tiny bit of firmness. Becky, in our zoo wolf pack, is certainly Alpha wolf, while I remain Beta wolf—at least as long as I do things the way Gray Boy likes them done. If I fail to live up to his standards, I'm promptly relegated to the bottom of the wolf hierarchy.

Yet even Becky had a problem retaining her status on one occasion.

When boy wolves approach sexual maturity at two years of age, they aggressively attempt to establish their position in the pack. Gray Boy gave his first indication of his try for power by running two of the keepers out of the wolf glen when they attempted to retrieve some of his precious food pans he had strewn about.

Encouraged by their hasty departure, he attempted to dominate my son Joel. Joel was in the glen attempting to repair the electric fence when Gray Boy came up to him. Jumping up, the big wolf placed his paws on Joel's shoulders. This was natural enough since Joel and Gray Boy Wolf had been good friends for a long time. Then, with his nose inches away from Joel's face, he suddenly snarled and bared his long fangs. Joel did the only thing he could do, safely. He simply froze. My other son, Rob, analyzed the situation instantly, picked up a clod of dirt, and threw it at the wolf to distract him. Gray Boy dropped down, backed off, and Joel made a hasty exit.

Gray Boy's action was understandable, however, in these two instances. Neither the two keepers nor Joel were actually members of the pack. The zoo wolf pack at that time consisted of Gray Boy, Akela, their younger brother Taurus, Becky, and myself. Gray Boy had, of course, already established dominance over Akela and he often made life miserable for her by snarling and throwing his body over her in the act of dominance, while she did the tummy-up submission posture beneath him. Dominating Taurus was no challenge, either, as far as Gray Boy was concerned. He simply ignored him.

Gray Boy attempted his next act of dominance only after long and careful consideration, I'm sure. Becky and I were in the den area with the three wolves, throwing sticks for them, playing with them, and feeding them monkey biscuits, which they dearly loved. Becky turned to feed Akela and, suddenly, Gray Boy snarled and leaped for the back of her neck. He had no intention of actually harming her. It was just a dominance attack, but it frightened me terribly. I immediately grabbed the big bully by the back of his neck, pulled him over backward on the ground, and uttered my infamous wolf howl. Instantly, Gray Boy became the picture of complete submission. He whined, wagged his tail, and licked my hands. While I held him there, Becky scolded him.

"Gray Boy," she demanded to know, "what is the matter with you? Shame on you, Gray Boy, shame. After all, I'm virtually your mother. If you ever forget your manners like that again we'll make a wolfskin muff out of you. Now you remember that."

I released the wolf then. Gray Boy walked away, head low, tail between his legs. His play for power had failed miserably. Gray Boy was in disgrace and he knew it. Becky and I turned our backs on him, walked out of the cage slowly, heads high, just to show Mr. Gray Boy Wolf that we were still Alpha and Beta wolves in the zoo pack.

When we visited our wolf friends the next day, I confess we did so with no little feeling of trepidation. We really didn't know what reception we might receive from Mr. Gray Boy. We had no sooner approached the gate, however, than Gray Boy immediately did his tummy-up act. He was, thank heaven, all submission and good will. We entered the area, scratched his tummy, and played with the three wolves for a long time. Never once did Gray Boy display the slightest sign of aggression or hostility. He had, apparently, learned his lesson —that Becky and I are the leaders of the pack.

We're the exceptions, however. Just recently I asked Robin Taha, one of our keepers, how Gray Boy was behaving. He had just finished placing the wolves' food in the den, after first dropping the doors to keep the wolves in the glen area. Taha looked at me for a long minute, then shook his head in dismay. "That gray wolf is simply ferocious," he told me. "He stands next to the fence while I open the den door, snarls, paws at the fence, even bites it trying to get at me. Frankly, he scares me to death."

* * *

For a long time Gray Boy gave Becky and me the respect we deserved as leaders of the zoo wolf pack. Yet every wolf—like the proverbial dog—will have his day. Just after the season opened in the spring of 1978, we had to take the two wolves out of their glen in order to treat Akela for an eye infection. Becky and I had put the leashes on them just a week before and taken them for a good run over the whole zoo so the city electrician could repair the electric fence in their exhibit area. The four of us had enjoyed ourselves immensely. Gray Boy was cooperative, enthusiastic and on his best behavior. Now we were in for quite a different experience.

After calling the wolves into the den, Becky attached the leash to Akela while I did the same with Gray Boy. We opened the den door and everyone rushed joyously out into the main zoo area. Just twenty feet away, Akela brushed against Gray Boy and he snarled and snapped at her. This was my first indication that things weren't going right. Yet when the panting, happy wolf pack arrived at the far end of the zoo and Becky approached Gray Boy and me, Gray Boy did an immediate, submissive tummy-up and permitted her to scratch and pet him.

Becky and one of the zoo crew applied some ointment to Akela's eye and we started our return to the wolf area. Then, for some unknown reason, Gray Boy went berserk. He began running hard, charging at some of the other zoo attendants who were close, and threatening them with very earnest snarls, growls, and an impressive display of his long white fangs. Inasmuch as I was on the other end of the leash, I was experiencing no little difficulty in keeping up with him. In fact, he literally dragged me over most of the distance. I was able to keep him from reaching and biting portions of our valuable zoo people by digging in my heels at the last moment, pulling frantically on the chain, and commanding him to turn right or left. Somewhat to my surprise, he obeyed my shouted commands.

Becky, in the meantime, had become justifiably concerned about her aging father. She gave Akela's leash to one of the zoo attendants to hold and raced across the zoo to where Gray Boy was doing battle. As she neared, Gray Boy turned and charged at her. I was able to pull him back just before he reached her.

There was very little time for words. "Becky," I gasped, "please go away. You're only making things worse. As you can see, he's not challenging me."

"Dad, you won't be able to get him back in the den without some help," she exclaimed. "He'll tear you to pieces."

Again she approached us. Gray Boy was facing her, watching, waiting. As she came closer he charged, snarling viciously. Once more I managed to pull him back and this time I did the only thing I could do. I grabbed him by the muzzle and forced him into the submissive posture. Although I was only able to hold him for a short minute, he obeyed my command to "come" when he regained his feet and started, feet flying, toward the other end of the zoo. I uttered my famous wolf howl and Gray Boy turned toward me, tail low and smiling, and brushed against me affectionately. Then, as if to say, "It's our world, brother wolf, let's go eat it up," he turned, in the wrong direction once more, and began dragging me toward happier things.

Becky ran in to help me with the heavy leash. It was instantly apparent that Gray Boy resented her presence. Turning, he charged at her, and this time I wasn't able to restrain him. Rising up on his hind feet, snarling, he bit her on the shoulder. I pulled him back, muzzled him with my hand, and again forced him into a submissive posture.

"Come, Gray Boy, come," I screamed at him. And Gray Boy did come this time, in my direction. As we neared the den I commanded him to "sit" and, when he complied (to my eternal relief), I shortened the length on his chain and petted him. Once I had established some semblance of control, I led him into the den and removed the leash from his neck. Gray Boy immediately rushed from the den into the open part of the wolf glen.

Perspiring profusely and shaking like a leaf, I walked over to where some of the zoo crew were gathered about Becky, examining her shoulder. To my surprise she wasn't even bleeding. Gray Boy had inflicted an ugly pressure wound with his front teeth. Despite his terrible growls and aggressiveness, he had only given her a warning nip. He had not used his canine teeth, though he could easily have done so. He could also easily have torn her throat open. Gray Boy, however, had chosen only to warn her, to challenge her position in the wolf pack. Needless to say, our three-year-old male wolf had proven his point. He had upped his position in the pack hierarchy to that of Beta wolf, at least. Although he has fully respected my new position as Alpha wolf in the brotherhood, I'm quite sure that I'm going to have to prove my dominance again in the not too distant future.

We lowered the door to the outside area and led Akela back into the den. Becky removed the leash and, once the wolves were safely established in their quarters, the zoo staff gathered at the refreshment stand for a cool drink and a bit of relaxation.

There we had the opportunity to compare notes and everyone agreed on just how frightening the snarls, teeth, and savage aggressiveness of a big wolf can be. Especially to the human on the other end of the leash.

CHAPTER 17

Hatari—Danger, Danger

A friend of mine once asked me a disturbing question. Does the staff at the Des Moines Zoo have more accidents and injuries than the personnel at other zoos? I had to think about it for a long minute before I could give him an answer.

No, I told him, I don't think so, at least not relatively. Other zoos have their problems and their casualties, too. Some of them have even more than we do. But, I had to admit, we have some unique problems in our zoo that make things a little more difficult and dangerous than many of the other zoos. At our zoo we don't just put our big cats, elephant, and reptiles into permanent cages and allow them to grow old and die there. We move most of our entire collection at least twice a year and, when you handle and move animals that often, I explained, there's always a possibility of trouble.

Snakes, I confessed, are always something of a hazard. It's my strong personal belief, though, that no animal collection can call itself a zoo unless it includes reptiles. Television has conditioned the children to expect a wide variety of reptiles in a zoo. We are careful, I told my friend, and have developed a technique in handling poisonous reptiles that eliminates a great deal of the risk. After all, I concluded, we have tube-fed and otherwise handled our "hot" snakes thousands and thousands of times and have only been bitten six times. I feel this is a pretty good percentage.

Virtually all of the zoos close to ours have experienced a serious accident or two, I told him. One director had an elephant fall on him while he was training her, and broke his leg. On another occasion, a golden eagle grabbed his leg and they had to use steel surveying rods to pry the big bird loose.

The director of a nearby zoo lost part of his finger to an enraged orangutan. Another zoo director I know was attacked by a big male chimp and barely escaped with his life. He was bitten badly on the leg. A chimp in yet another area zoo apparently made a practice of biting off fingers from keepers and visitors alike. A keeper at this same zoo had his arm severely mutilated by a leopard that reached out and grabbed him as he passed her cage.

Just a year ago, I told my friend, a zoo keeper in Africa neglected to lock a safety door and a lioness escaped. The cat killed two children. In another recent incident, a tiger escaped from a zoo cage and, when the director attempted to force him back in, the cat attacked and killed him.

At a zoo just 150 miles from ours, a tiger escaped from a moated area and mauled a woman and child. A polar bear in a zoo 250 miles north of us mauled a man who fell into his moat.

The December 26, 1975, issue of the *Des Moines Register* carried the story of a woman who was attacked by a lion at a drive-through safari exhibit in the South. She was saved by an attendant who beat off the animal with a rake. The story went on to say that about a dozen visitors and several park employees had been attacked and injured since the park opened in 1969. One man was gored to death by an African Cape buffalo.

Poisonous snake bites among zoo people who handle reptiles are so common it is difficult even to recall specific instances. As Dr. Sherman Minton once stated so well, it's never difficult to tell zoo herpetologists from the museum reptile people at some of their mutual conventions. Most of the zoo people have missing fingers from reptile bites.

All in all, I feel we have done pretty well. Particularly since we attempt things that most zoos do not, such as our reptile lectures, venom-milking demonstrations, cobra charming, and affection-training with our big cats and other predators. Most probably, I continued, the other zoos don't have to engage in such activities because they have already become established and can offer their animals adequate facilities. They can provide nice natural areas for their animals, while we still exist in a state of limbo. We have to do something different to engender enough interest and support for our little zoo and in the very near future provide large, open areas for them. If we fail to do this, we will simply be closed down. It is as simple as that, I told him. I think he got the message.

CHAPTER 18

Humans Bite Too

Nineteen seventy-five began as a disaster, and the exhibit season ended as a catastrophe when our affection-training program became a matter of controversy. The results gravely affected our zoo and will for years to come.

Two members of our Zoo Association's board of directors, both men, objected to our practice of hand-raising animals and working with them in the manner we had for years. I attended the meeting when the matter first came up, but I was too speechless, too astounded, to discuss the problem with them.

Not everyone agreed with them. Some of the women present expressed the idea that our animals even benefited from our affection-training. One of the women had read Hediger's book *Wild Animals in Captivity* and enthusiastically supported our endeavors

The question raised is undoubtedly one of the most basic problems that exist in the zoo world today—how wild animals are to be treated in captivity.

A majority of the zoos in this country still keep their predator animals in cages. Most zoo people agree that cages will someday be a thing of the past and that animals will be confined in large, open areas surrounded by moats or fencing. Until that day comes, however, cages present problems in animal management and animal welfare that must be solved.

Some zoo directors still believe that their big cats are happy and have ample exercise room in a ten-foot by twenty-foot cage or smaller. They insist that the animal regards its cage as a sanctuary and home which it would be reluctant to leave if it could. They go even further and insist that while a lion may look lazy and listless to the zoo visitor, it is a negative anthropomorphism to suggest that the animal is unhappy there, needs more room, or may be bored. They base their arguments on the observations of field researchers who have studied lions in the wild and have found that they spend a great deal of time just lying around in the heat of the day. They neglect to mention that those lions undeniably exert themselves greatly during the night hours in the search for food.

Our observations, after affection-training predator animals for nine years at the Des Moines Zoo, lead us to a different conclusion. I'm convinced that the only reason the usual zoo lion or tiger would be reluctant to leave his little cage is that he's just too afraid of what lies on the outside. Our affection-trained animals can't wait until we get the leashes on their collars, the cage door is opened, and they can run, romp, and play in the grass.

My supporters agreed that we had little choice in our method of working with the animals. They agreed it was much more desirable to lead an animal over the distance between the winter quarters and the exhibit area, to drive him in my station wagon, or to coax him into a transfer cage, than to force him into a cage with prods, water from a hose, or a shot from a tranquilizer gun.

One of the women on the Association's board of directors told me, however, that I had a bigger problem coming up. The Zoo Association had engaged a consulting firm to help with our expansion plans, and one of the consultants was vehemently against hand-raising (as they erroneously called our affection-training techniques) animals. "He," she said, "considers it undignified to handle animals in zoos, to take them from their mothers when they're young and put them in petting areas, or to give performances with them."

This made me quite angry. "After all," I pointed out, "ninety percent of our affection-trained animals are born at the zoo, in dens that don't have enough heat in them for little animals and in weather that is either too cold or wet for them to survive. What am I supposed to do with them? Should I just let them perish?"

I paused for breath. "We have," I continued my argument, "received some young tigers, leopards, and jaguars over the years that now make up our breeding stock. All of them arrived singly, excepting Ramses and Nefer, our big lions. All of them were just cubs, too, because it's much easier and much less expensive to ship a baby lion or tiger than it is a five-hundred-pound adult one. We had to work with those young animals," I insisted, "if we were to move them back and forth between the winter quarters and the exhibit area two or three times a year. And, as you know, these cats have been the greatest attraction at our zoo, for years. Without them we could not have remained open, with our sad little exhibit areas."

"Oh, I agree with you," she said. "Don't eat me up. I've read both of Hediger's books on keeping wild animals in captivity and I agree with him, and you, one hundred percent."

But I was all wound up by now. "Why must any animal remain in a constant state of hypertension in captivity?" I demanded to know. "They certainly don't live that way in their native habitat, as many animal studies have pointed out. Why do people think that the normal emotional state of a big cat in captivity should be one of constant snarling aggression? Any human or animal psychologist could tell them in a minute that a snarling, aggressive animal is simply a frightened animal. And this is true of an animal that evidences cage boredom or withdrawal symptoms. The basic factor is fear; the expression depends upon the animal's individual dynamics or the particular time or circumstance."

"Oh, come on, Bob," she replied, "calm down. It's not all that bad. Most of the women on the board support your view. The consultant brought out another point, though. He said that some zoos hope to eventually return their big cats to their native lands and natural habitat. That's another reason, he feels, that the animals should not identify too closely with a trainer, or humans in general."

"Good Lord," I groaned in utter frustration. "That guy should talk to any falconer. Falconers have taken birds from the nest for thousands of years and trained them. Quite often the birds escape or the falconer sets them free. The birds survive because they know how to hunt for themselves. The falconer has cared for them and trained them until they know how to capture their food. The same thing was true of the lion, Elsa, in *Born Free*. Mrs. Adamson worked with her lion for months, feeding it often after she had released it, until it learned to hunt for itself. Any zoo director who thinks an animal can successfully hunt for itself simply because it snarls and tries to eat the keepers has better delve into animal psych a little deeper."

"I suppose you're right," my friend said thoughtfully. "After all, their parents teach them to hunt in the wild, and it would be pretty difficult for humans to teach them to hunt if the tiger, for instance, just didn't like humans."

"That's about right," I agreed. "The tiger would have only one advantage. He wouldn't have to look far for his meal. He'd just eat the first human that was handy."

"Well, I'm behind you. I want you to know that. Even Peter Batten, who seemed to find very little that pleased him about the zoos in this country, stated in his book *Living Trophies*, that he agreed one hundred percent with Hediger that tameness is very, very desirable in captive animals for a great many reasons."

I nodded in agreement. "That's probably the only thing that Peter Batten and I will ever agree on, but I think he's right there. Actually, Jacque, I believe the day will come when every zoo will have an animal psychologist on the staff, just as they do a veterinarian. His sole concern will be the emotional well-being of the animals. Some zoos are attempting something of this sort now with structured exhibits that permit the animals to climb, leap, and run, but these physical changes in exhibit areas are far from being the total answer."

I told my friend how much I appreciated her thoughtfulness in warning me of my impending problems. She had made a special trip to the zoo just to keep me informed and I was very grateful. "I'll give the matter a lot of thought, Jacque," I promised her. "Perhaps I can figure out a happy compromise that will please everyone."

Jacque, in turn, promised she would help me in every way possible.

The more I thought about the criticism from the two board members and the consultant, however, the more concerned I became. I invited the officers of our local Audubon chapter and the president of a humane society to visit our zoo, investigate our affection-training program, and evaluate it for themselves.

It wasn't difficult to demonstrate the differences between our affection-trained big cats and those that were not affection-trained. We had three leopards that we had not been able to affection-train simply because we lacked the necessary time and personnel when they were young animals. Two of the cats were two years old; one was four. All were mature. When I approached the cage of the two younger leopards in the exhibit area, their response was instant, snarling aggression. When I went closer to the bars, they made every attempt to attack and kill. I explained to the visitors that two of my men had received claw wounds from one of the cats while feeding them through the safety feed door. They were obviously a constant hazard.

We next visited the winter quarters where the older leopard evidenced the same determined desire to attack us all, since we were grouped about his cage. To show the difference in behavior dynamics, I walked down a few cages to where Shannon Leopard was housed. Shannon is a people leopard. She pressed up against the bars, asking me to tickle her nose with my finger. She was obviously happy to see me.

Our next stop was Bagheera's cage. Black leopards have a popular reputation for being very aggressive and dangerous. Not so Bagheera. The Seneca Zoo in New York had given him to us when he was just over a year and a half old. They had done a wonderful job of raising him. I told my investigating committee that when I had picked him up at the airport Bagheera was sound asleep in his crate. Certainly he hadn't undergone any of the terrible stress many animals experience when transported from one zoo to another. And, as the curator of the Seneca Zoo had warned me he might, Bagheera jumped right into my arms once his shipping crate was opened.

We visited the rest of our affection-trained animals: our big lions, Brucie Tiger, Mercury Cougar, the jaguars, the wolves, and our wonderful Leonardo Lion. They all came to the front of the cage as we approached. They were friendly, alert, and happy to see us. There was none of the aggressive hostility evidenced by the three leopards.

"Do the animals look undignified?" I asked the group.

"No, nothing of the sort," one of the visitors said. "I think they look as though they're well cared for, and they're obviously more at ease in their cages than the leopards are." All the others nodded in agreement.

"Do other zoos affection-train their animals?" one of them asked.

"Not that I'm aware of," I confessed. "Many of them take the cubs from their mothers and hand-raise them for petting, or contact, areas, but none of them, to my knowledge, spends the endless hours, days, and months of effort required to affection-train an animal. The process requires a year of difficult work," I explained, "endless patience and some deep insights into predator-animal behavior and psychology. It should only be attempted by persons who have had a great deal of experience in training animals."

I paused and thought for a moment. "About the only comparison I can think of is with the animals that are used in TV commercials, motion pictures, and TV programs where tame animals are necessary. These are in a sense affection-trained but, from what I've been told, virtually all of them have been declawed, defanged, neutered, or altered in some other reprehensible way. Any trainer that would mutilate an animal in such a way just to make money is too despicable to talk about. I simply turn the darn commercial or program off."

I took Leonardo Lion out for a walk and permitted him to play in the grass with my son, Joel. The Audubon people were delighted with the huge cat and his obvious love for humans. I explained to them that I did not consider this an act or a program. The main purpose was to give the lion the needed exercise his cage wouldn't permit. Of course, I confessed, we took advantage of the crowds that gathered to watch the big cats walk about on the leash and play and romp in their exercise area. We simply stuffed the spectators with information about the increasingly important place zoos had in preserving hundreds of endangered species and protecting them for the future.

The Audubon Society officers and the president of the Humane Society were wonderfully enthusiastic about our affection-training. Despite all the time and effort it entailed, they felt the results were most rewarding. The only problem, they said, was still the matter of the cages. We all agreed on this. Although our big cats were at ease and lived without constant fear in their twenty-by-twenty-five-foot cages, they certainly needed more exercise room.

Despite their encouragement, though, I turned chicken. After thinking it over, I decided to capitulate rather than risk an all-out war with my critics in the Zoo Association and the representative from the consulting firm.

I concluded that our special activities must go. We would continue with a portion of our reptile lecture, but I wouldn't work with poisonous snakes or charm cobras. A bite at this inopportune time, I felt, might create controversy at the very moment we were attempting to work with the consultants on the zoo's expansion. Controversy I didn't want.

We decided to discontinue our very popular Adventure Activities for the time being as well. After all, if a child fell from the back of our little elephant or one of the burros, this too might provoke criticism about the zoo. Of course, there was a negative aspect to all this that was pretty plain to me. And I was right—our attendance dropped 25 percent for the season.

We continued to work with our big cats though. I just didn't have the heart to leave them in their cages. Every day we took those we were affection-training out for their exercise and play. We just didn't do it when there were zoo visitors around. Becky Elgin still walked her two young wolves every evening and continued with their obedience training. We continued to work with the elephant, too, for certainly none of us wished to carry her back to the winter quarters once the season was over. We just didn't give free elephant rides any longer.

We added a new activity that summer, however. Our famous Snake Cult gave two performances and many people wished we had given them earlier. This colorful little pageant has been one of the most popular educational programs we've developed at the zoo. Our girl guides are costumed to depict some of the famous personalities that have been associated with reptiles down through history. Eve, for instance, appears on stage wearing a very big fig leaf and carrying a boa constrictor. The narrator tells the audience some interesting facts about the first man and woman and their tempter, the snake. Medusa follows next, carrying a red rat snake. Our narrator relates the legend of Medusa. Next comes Cleopatra and her story. The pageant includes characters, costumes, and interesting anecdotes from the beginning of man's first association with reptiles through Greek legend, Egyptian history, the snake cults in Africa and India, to the cowboys and their problems with rattlers. We conclude the show with an explanation of the present-day snake cults in the southeastern part of the United States and how they manage to "take up serpents."

As I mentioned before, we gave two performances, but we didn't give them just because we hoped to increase our attendance. We did it for a more desperate reason. Iowa experienced a prolonged drought during the summer of 1975. It failed to rain for weeks and it was very hot as well. We decided to try to help by emulating the Hopi Indians and conducted a rain dance—or at least our kooky modern version of it. Of course we invited the news people to witness it. We were confident of success.

The cast of girl guides donned their cute little costumes, took up their harmless serpents, and began to dance in a big circle. A tom-tom beat out the solemn tempo. I entered the circle carrying one of the zoo's tame rattlesnakes. We then all knelt down, lifted our snakes to the sky, and yelled, "Rain, rain, rain!"

Of course it rained—that very night. Not a big rain, but certainly a very welcome one.

One of the news people suggested, after this success, that we try again. A couple of weeks later we gave another performance on our weekly TV show. Soon afterward it rained six inches! This was a new record for Des Moines and vicinity. We gave up after that—we didn't dare turn on the spigot again and maybe cause a flood.

* * *

Perhaps the snakes at the zoo resented being exploited in this manner, for they started picking on me again. Huff Cobra, the big Asiatic who had almost killed me in 1969, apparently wanted the last word. We were posing for pictures for a photographer and I was holding Huff up, high and hooded, so the photographer could shoot him in this unusual position. It was different and it was dramatic. Huff was posing well, almost oblivious to my hands and my presence. This is not too difficult to accomplish, really. The cobra's attention was fixed on the camera and the man who was holding it. Since cobras have one-track minds, the charmer can often hold the reptile in this manner almost indefinitely, or at least as long as the snake's attention is concentrated on something else.

As I certainly should have known, however, Huff is not the usual predictable cobra. After three or four minutes he began to tire of the photography bit and turned his big, hooded head toward me. I immediately attempted to drop him into his snake box.

Huff was too fast for me. He looked at the box, then again turned my way. He then, deliberately, bit me on the right hand. Blood streamed from the little hole. Finally I managed to throw him into his box and we closed the lid.

Moments later the pain hit hard. Huff had injected venom. He had not, however, held on and chewed as he had in 1969, so I had some reason to believe I hadn't received a large amount of toxin.

The photographer was decidedly upset by the incident. "Don't you think you'd better get to a hospital as fast as possible?" he asked. "I've got enough photos, I'm sure, so let's call it quits and get you on your way, right now."

"Oh, I don't think I've got much to worry about," I told him. "I've got plenty of immunity, I'm sure. I'll be glad to get out another snake if you think you need to shoot some more pictures."

"Good heavens, man," he exclaimed, "didn't one of these things nearly kill you a few years ago?"

"Yes, that's true," I replied, "but I've been bitten by two other cobras since then and didn't experience anything more than the usual pain. Nothing serious happened at all."

He looked at my hand. It was covered with blood by now. "No," he muttered, shaking his head, "let's call it off. I don't need any more pictures, and I sure as heck don't want to see you get bitten again. Actually, it makes me a little sick." He gathered up his equipment, told me he would have the photos ready in a day or two, and departed.

I walked down to the kitchen and washed the blood from my hand with some antiseptic soap. My hand was beginning to swell and the pain, as always, was intense. I decided to drive home for a little rest and some aspirin.

My wife met me at the door. Someone from the zoo had called her and told her I had been bitten again. "We're going to the hospital, right now," she declared, very firmly. "Get back in the car and let's get started."

"Let me sit down for a while, will you?" I pleaded. "I'm kind of tired and would like to relax a bit first."

"No way!" she stated. "You're not going to pull anything like the trick you did when that second cobra bit you."

I winced at the thought. That, I'm sure, I will never be permitted to forget as long as I live. Two years earlier I had been bitten by a small cobra when I was treating him for a burn wound he had received from a heat lamp. It had been on a Monday when the zoo was closed. Earl Connett and I were the only ones in the exhibit area when it had happened. Jane and the younger kids were visiting the state fair. Knowing the anguish and horror she had gone through when Huff had nearly killed me in 1969, I decided I would spare her the worry about the cobra bite. I had convinced Earl that the cobra was, after all, rather small and that I had plenty of antibodies to neutralize the toxin. I had made him promise that he wouldn't call Jane under any circumstance. He had reluctantly agreed.

I had then driven down to the hospital and haunted the waiting room adjacent to the intensive care ward, just in case I had made an error in judgment. I spent the whole day there, in an uneasy state of mind, I confess, but knowing that if the cobra venom did overcome my immunity they could get me on the respirator instantly and begin treatment with my human antivenin, which was frozen and stored just across the street from the hospital in the Blood Bank.

Nothing had happened after twelve hours, so I drove home. My wife was furious when she heard the whole story about why my hand was so badly swollen. She yelled at me and stalked about, hissing and spitting words like a lady leopard. I cowered on the sofa in the corner and promised that, yes, I would let her know immediately if anything of the sort happened in the future.

Now I had no defense. There was no arguing with her. "Let's go," she insisted, pulling at my shirt. "Come on, Bob, get in the car."

"We'll have to sit in the waiting room for the entire night," I warned her, "because I'm not about to call one of the doctors and have him admit me to the hospital. The minute I do that the reporters will hear about it, and heaven knows I don't need any adverse publicity right now in view of the problems I'm having with the Zoo Association."

"I don't care how long we wait," she answered. "I'd rather wait down there than sit here and worry about you all night."

We gathered up a number of vials of my lyophilized blood serum to take with us. While I had faith in the circulating antibodies in my blood, I decided to take a supply of my human antivenin along as well, just in case.

Jane and I arrived at the hospital and explained the situation to the nurses. They readily agreed that we might spend the night in the visitors' lounge and placed my supply of human antivenin in the refrigerator for safekeeping.

We made ourselves comfortable and prepared to wait it out. My hand was slightly swollen and painful and we watched it carefully to see if I might experience anything more than local involvement. Hour after hour we sat there, visiting with other people, drinking soft drinks and coffee. Both of us became very sleepy. We waited and watched my hand for further symptoms. Nothing happened.

When morning came, the swelling in my hand had receded and I had no signs of neurotoxic poisoning. We thanked the nurses for their kindness, picked up my antivenin, and departed for home. Once there, we crept into bed and slept the entire day.

Once again my self-immunity had proven its value. And once again, in a sense, Bill Haast had probably saved my life for, after all, he had pioneered the whole process of immunity buildup and had helped me in my efforts to achieve protection against cobra venom.

CHAPTER 19

Leonardo, King of Lions

If the 1975 exhibit season was a bust as far as attendance was concerned, if the criticism heaped upon our affection-training program rankled and irritated me greatly, if the snakes again did their best to do me in, the season was still a great joy and success as far as the Elgin family was concerned. We, and the zoo, had Leonardo Lion and that was a beautiful experience all in itself.

Leonardo Lion is, in my mind, the king of all lions. Even as a little cub he was very unusual. He was born in February of 1974, just after Skipper Chimpanzee had bitten off my fingers. He was one of a litter of three and, as usual, it was too cold in the den for the babies. We got them out as soon as possible, but they developed the usual chronic pneumonia. We treated them with everything in the medical world and managed to save their little spirits.

They graduated from the incubator to our bathtub, just as twenty-one of their previous brothers and sisters had done, and soon made the Elgin home into a big lion den. They were still on the bottle and my wife was responsible for feeding and caring for them. I was unable to do a thing. My hand was most painful and still in bandages.

Lions, like humans, are individuals. Every one of the cubs we have raised and affection-trained possessed a different personality. Janie Lion, for instance, was always shy. Every time a stranger came into the house she ran and hid. Becky Lion was confident and aggressive. Little Luff Lion, the one we're training now, is a bundle of energy. Ramses, the sire of all these cubs, is one of my favorites. We could always do everything with him—except touch his head. Nefer, the mother lion of all, has always been open, friendly, and gentle. Leonardo, I believe, is very close to being the perfect lion, if anything can be perfect. As a baby he was calm and collected. Noises never bothered him. He had poise and confidence and lots of class. He enjoyed strangers and loved to ride in the car. Above all, he adored my wife, Jane.

The two other cubs were sent to homes in other zoos when they were three months old. We simply couldn't bear to part with Leonardo.

He became something of a permanent house guest and stayed with us until the zoo opened in June.

Leonardo made several visits on the zoo's TV show that season and became something of a local celebrity. We even made an exception with him at the zoo and occasionally permitted small children to pet him while Jane held him on her lap. We no longer place exotic animals (particularly baby predators) in our petting area, because we've learned these animals don't endure the rough treatment too well. Domestic goats, lambs, rabbits, and baby burros seem to stand the stress of children playing with them without too much difficulty, but even a pat on the head from the wrong angle or at the wrong time may ruin a baby lion's or tiger's relationship with humans for the rest of his life.

We had some unusual experiences with Leonardo Lion. Most of them were delightful, but one that took place in late May of 1975, just before the zoo opened for the season, provided us with a unique insight into lion personality and again proved, in a dramatic manner, that big cats should never be kept in homes.

Some of the news people had expressed an interest in our affection-training research so we arranged for a photographer to shoot some pictures of Leonardo playing with Jane. Leonardo was eighteen months old and weighed 375 pounds. He simply towered over Jane when he stood straight up beside her and we thought this contrast in size would make an interesting news photo.

We had only one nagging problem about the matter. Leonardo had been in the winter quarters all winter long—except for an occasional outing—and he had not seen Jane for at least six months. I'm quite fond of my wife, of course, and I was concerned about how Leonardo might conduct himself after not seeing her for such a long time.

As a precautionary measure, we decided first to chain Leonardo in the area just to the side of the winter quarters building. We accomplished this and I was playing with the happy Leonardo when, as we had planned, Jane left the building and slowly approached the two of us. Leonardo stopped his play, sat down, and watched with great intentness as she came nearer. He just could not believe his eyes.

Afraid that he might knock her down if he made an enthusiastic charge once she was within the limit of the chain, I held him back by the collar. Or I tried to. They met halfway. Leonardo was simply overjoyed to see his mistress. He went wild in a careful, controlled sort of way. He jumped up, placed his huge paws on her shoulders, licked her face and hair, pressed his great body against her, and moaned with pleasure and love.

No dog, domestic cat, or any other animal ever evidenced a greater love for a human than Leonardo did that day. Twice the sheer weight of his body knocked Jane off her feet, but never once did Leonardo even place a paw on her while she was on the ground. It was obvious that the big cat was being most careful not to hurt her in the slightest.

I was concerned, though, after a while, that he might unintentionally do just that, so I tried to call him over to me. Leonardo ignored me completely. This disturbed me a little and, since the photographer had succeeded in getting the pictures we needed, I decided it was now time to put Mr. Lion back in his cage. Again I tried to call him away from my wife. I "uuuurfed" at him, in lion language, two or three more times. Leonardo wanted no part of it. He ignored me again.

With the knowledge that no lion can resist this invitation to run over and play, I got down on my hands and knees and "uuurfed" at him. For a moment he hesitated, then he walked over to me–slowly, stiff-legged, ears back. As he neared he raised one big paw and hit me on the head like a sledge hammer. I saw a million stars and thought for a moment that he might have broken my neck.

Leonardo turned his back on me, again went over to Jane, and rubbed his long body against her legs with all the love in the world.

I realized we had a real problem. We had a very jealous lion on our hands, a very big lion, and I wasn't sure just how to get him away from my wife. I had seen something of the same possessiveness between our parent lions, Ramses and Nefer, and our big wolves, but this was the first time I had ever heard of such a psych dynamic between a cat and a human. Leonardo liked me, but he loved my wife and he resented my presence. It was as simple as that.

He permitted me to approach and take him by the collar. This I did with great care and caution. While I held him under some semblance of control, Jane and I walked him the full length of the chain, and Jane was then able to walk away from him. He watched her, sitting completely motionless, as she disappeared around the side of the building.

He waited for her to reappear without moving a muscle. Then, apparently resigned to her leaving, he turned his attention to me. We played for a moment while I unsnapped his short, heavy leash from the long chain, then turned to go back to his cage in the second winter building.

In the meantime Jane had walked completely around the office building and was waiting for us about a hundred feet away. It had been my idea. Leonardo usually preferred to remain outside and play rather than return to his cage. I had felt that Jane could help lure him back inside, so I had suggested earlier that she stand in that place when we started to lead him back to the building. I had certainly reconsidered the arrangement, in my own mind, after witnessing Leonardo's profound love and jealousy, but I had forgotten to tell Jane not to get anywhere near the cat and to please stay out of sight.

Leonardo did everything I feared he might, and more. Seeing Jane, he of course wanted to play with her some more. He charged toward her. I tried to hold him back. No human holds a 375-pound lion back when the lion really wants to run. And lions can run very rapidly when they wish. Leonardo simply pulled me along as though I wasn't there. I somehow kept my feet. I knew it would be a disaster if I lost my balance, fell, and he turned on me.

On the other hand, I was afraid that he might knock Jane down at the speed he was going and hurt her. Just before he reached her, I dug my heels in, pulled with all my might, and managed to swerve him a little—just enough to permit me to get between him and Jane.

Leonardo didn't like that. He growled, seized my forearm in his big mouth, and shook me like a rat. "No bite! No bite!" I screamed at him. I bopped him on the nose with my free hand. The cat dropped my arm and backed off. Then Leonardo gave that awful warning cough of a lion who is very displeased and tried to attack my legs with his claws. "No bite, Leonardo!" I screamed at him again. Once again he backed off, snarling.

An instant later he rushed me again, claws out, trying to rake my legs and pull me down. And again I backed him off with the command. He walked over to Jane, crouched down beside her, and looked at me unhappily. I was far from happy about the situation myself, but obviously I still had the cat under some semblance of control.

Leonardo grumbled his displeasure, wanting me to go away and permit him to play with Jane a while longer.

I was thinking fast and hard. "Jane," I told her, "please, just walk away from him now, slowly and carefully."

"Absolutely not," she replied. "I'm not going to leave you here to be killed. If I start to walk away and Leonardo follows me and you try to stop him again, he may try to kill you. You know that."

I realized she was probably right. Then she came up with a different idea. "Why don't you give me the leash," she suggested, "and let me lead him into the building."

I was absolutely horrified. "Good Lord, woman," I yelled at her, "I'm not about to give you an eight-foot leash with an angry lion at the other end. What kind of husband do you think I am? What happens, for instance, if you manage to get him inside and he refuses to get into his cage? You can't pick him up and carry him in, after all."

Neither answer was the correct solution, that was apparent. We were in something of a dilemma, but fortunately we had prepared and practiced for just such an emergency. I yelled for one of the keepers to undo the long chain we had used to tether the cat for the photographs. The keeper raced over to the site, unlocked the chain from the stake, then ran over to where he could safely throw the end of the chain to me. I secured the snap to the leash and told the keeper to fasten the other end to a telephone pole fifty feet away.

"Now, Jane," I told my wife, "let's both back away from the cat at the same time. Do it slowly. You go toward the fence and I'll go toward the building."

Leonardo watched us curiously, undecided as to what to do. The chain was at full length, and when he tried to follow Jane he found it impossible. By that time, I, too, was out of his reach. Leonardo sat down and watched my wife until she disappeared behind the office building. And when she reappeared again, on the other side, he tried to run toward her once more. Thank heaven this time he had a huge telephone pole to stop him and not a fragile little two-hundred-pound human like me.

I walked over to the building where Jane was standing. "Are you all right?" she asked. "How's your blood pressure?"

Sometimes I think women are just innately braver than men. "Well, it's certainly better now," I said. "Thanks for not leaving. I do think that Leonardo would have eaten me if you had." Then I hugged her for a long minute.

I asked the keeper to take my wife home in the zoo truck and bring my son Joel and the tranquilizer gun back with him when he returned. I disliked having to tranquilize Leonardo, but I was still concerned about his emotional state and certainly had no desire to approach him again at the moment to try to lead him back into the building. I needed additional help and I intended to take every precaution.

The photographer and I waited while the keeper drove the truck to my home. We both kept a safe distance from Leonardo. The lion, for his part, just sat there watching the building where he had last seen Jane. He didn't move an inch. He was waiting patiently, sure she would reappear and play with him some more.

Minutes later the truck returned with Joel and the tranquilizer equipment. Joel had literally grown up with lions and other big cats and, along with my son Rob, was the person I always depended upon to help me train and work with them. He had been ill that afternoon with a bad cold. His mother had insisted that he stay in bed, so he had not been available to help at the beginning of our little fiasco.

Joel sized up the situation more objectively than I was capable of doing at the moment. He merely walked up to Leonardo and talked to him in a stern voice. "What in the devil have you been up to?" he demanded.

Leonardo blinked his eyes, "uuuurfed," and pressed against Joel's legs. Encouraged by this show of affection, I summoned up sufficient courage to approach the two. Leonardo jumped up, pressed his magnificent head against mine, and rubbed against my face. It was his way of showing he had forgiven me. Joel unsnapped the leash from the chain. We led our big lion to the building and into his cage.

We examined my arm after that. There were several black-and-blue marks but that was all. Leonardo had merely given me a warning. A warning to keep my distance and to permit him to play with Jane.

It had been an uncomfortable experience, but I learned a lesson from it: Never, never get between a lion and the woman he loves!

Once again, at the end of the season, we walked the animals back to the cages in the winter quarters. Never once were we forced to use the tranquilizer gun.

Leonardo was the last to leave the exhibit area. We wanted him to soak up all the fresh air and sunshine possible before we locked him away in his winter dungeon.

I fastened the leash to his collar and led him from the cage. He was most cooperative and happily jumped into the rear of the station wagon. We closed the window behind him and drove to the winter quarters. We then opened the back and Leonardo jumped out onto the ground.

I next tried to lead him, with the leash, through the big open door of the building. Leonardo made it most apparent that he would not enter that awful building again, under any circumstances. He decided to go in the opposite direction, and since he now weighed over four hundred pounds I went with him. Only this time I did not keep my feet. Leonardo pulled me through a big pile of straw and horse manure. He began to run, and I simply could not hold onto the leash. He pulled it out of my hands and we now had a problem. Leonardo was quite free to go anywhere he wished. If he began parading all over the south side of Des Moines, I knew the zoo would soon have a new director.

"Grab the leash," I yelled to everyone present, still prostrate on the ground.

That was hardly necessary. Leonardo was a nice lion. He merely stopped, sat down, and waited for Joel to walk over and pick up the end of the leash. Joel handed it to me, once I had regained my feet and some semblance of composure. Again I tried to lead Leonardo toward the building. This time the lion jumped into the rear of the station wagon. He felt, I'm sure, that he had outfoxed us.

An evil idea came to me. I smiled to myself, pleased at my cleverness. "There's more than one way to skin a cat," I told the others. "Close the rear window, Joel, and we'll take Mr. Leonardo for a little ride."

Getting into the driver's seat, I backed the vehicle through the door of the building until the rear half was completely inside. We again opened the back end and asked the reluctant Leonardo to come out. Leonardo refused. He crouched in the far corner, obviously determined to stay there. I "uuuurfed" at him, and he came over to the open end and rubbed his head against my face. Again he retreated to his corner. We threw some choice pieces of horsemeat on the floor. Leonardo ignored them. We then spun a big feeding pan around and around on the cement floor. Leonardo listened, looked and, though interested, decided to remain where he was.

Again I "uuurfed" at him. He walked to the back of the car where I was standing and again rubbed his head against my face. I tried to coax him down to the floor. He turned and walked back to his corner. We threw more meat down. We begged, we pleaded. All to no avail. Leonardo simply did not want to go back to his old cage life.

Then I tried another idea. "Joel," I said to my son, "get down on your hands and knees and crawl across the floor back here. Act like a little lion and maybe Leonardo will accept an invitation to jump out and play."

The floor was hardly clean. Joel, with some reluctance, got down on all fours and crept across the room. Leonardo found this very interesting indeed. He walked to the rear and peeked out at the strange would-be lion creeping about beneath him. Joel "uuuurfed" and moaned to him in lion language. Leonardo still hesitated.

"Roll over on your back and 'uuurf' at him," I suggested.

That did it. Leonardo leaped from the station wagon and landed like a big boulder on poor Joel's prostrate form. Fortunately, Joel is very big and absorbed the impact without damage. While he played with Leonardo for a moment, I closed the rear window on the station wagon so Leonardo could not jump back inside.

I grabbed the lion's leash and lured him away from Joel. We coaxed him over to the door that opened into the adjoining room where his cage was. He went through it reluctantly. We led him down the aisle to his cage and he stopped. We talked to him, we begged him please to go into his cage. He refused. More meat, this time inside the cage. Leonardo ignored the offering. Finally we did the only thing we could do.

Joel lay down on the floor, half inside and half outside the cage. Grabbing big Leonardo by the front feet and legs, he pulled away at that end of the cat while I pushed at Leonardo's rear portion with all the strength I had. We managed, between us, to get the reluctant lion inside and the door closed. Leonardo left his mark, though. There were two big, wet lion tracks on Joel's T-shirt where Leonardo had walked on him.

CHAPTER 20

The Nature of the Beast

Over the years, as Joel and Rob and I worked with our big cats and other predator friends, we came to appreciate fully just how wonderfully adaptable, gentle, and intelligent these supposed monsters really are. We learned to talk their language. The deep chest-tone grunt, "ooooough," that is the lion's greeting sound, and the high-pitched "rowwrr" of a lonesome little lion became part of our daily vocabulary. We also learned to cough, spit, bark, and snarl when it became necessary to use bad words in lion language.

We learned, too, the language of tigers with their soft "pffft" whisper of greeting and their long, sad "owrruuuu" of loneliness. We were able to understand and mimic the high "neeoww" pussycat cry of the leopards and jaguars when they wished to express friendship, and we talked to them in this submissive way when we approached so we would not frighten them.

As we learned to understand them better, we realized that when they became aggressive and threatened us they were not just "reverting to nature" as most people and even some zoo directors seem to think. We began to see that they became aggressive because they were afraid, as all predators often are, and that if we were really to become friends with them we must first overcome their fears as much as possible.

We learned, also, early in our endeavors, that such powerful creatures must have some direction in understanding their limitations as far as we were concerned. Only a measure of discipline could accomplish this. But it required just a small touch of discipline, judiciously applied. First, love had to be there in abundance to make discipline bearable, creative, and effective.

Many of these insights, such as body position, gestures, tone of voice, warning signs, and other preverbal communication between man and animal, came only from experience over the years. These are the essence of affection-training, and no one, to my knowledge, has written a "how to do it in one easy lesson" book on this relatively new relationship between man and predator animals.

* * *

Affection-training demonstrates, I believe, the highest possible degree of mutual love and trust between a predator animal and man. The relationship is greatly different from that achieved by a circus trainer and his animals. A circus lion, for instance, is taught above all to remain on his pedestal until he's cued to do his trick. Most lions in the act simply sit there and do little or nothing more. They are there for atmosphere. The trainer may "bounce" them by poking at them with the butt end of his whip, and the lion will snarl and strike at the trainer with his paw. But the trainer keeps a very safe distance and very, very few trainers bounce a cat more than twice. To bounce a lion more than twice in succession is considered a hazard to the trainer's health. The lions performing in the act go through their routines on cue and leap from one pedestal to another or jump through a hoop. Perhaps they also line up on the ground before the trainer and then all lie down together. Then comes the final "rise," when all the lions sit up straight on their pedestals with their front paws in the air. The door to the arena is opened by an attendant, the trainer stands, legs apart, in front of the exit, and the animals rush out between his legs and back to their waiting cages.

Working with an affection-trained lion is a very different matter. The physical contact between trainer and animal is infinitely closer than with circus performers. The animal walks beside, in front of, or behind the trainer on a very short leash. The cat may cuff at the trainer's legs, or he may simply jump up and wrestle him to the ground. Then the lion usually sits on the trainer's prostrate body and smiles in triumph.

Even more exciting, for the lion at least, is the little game of cat and mouse. The cat, Mr. Lion, is taken into a large exercise area, the lead chain is taken from his collar, and lion runs happily after mouse, Mr. Trainer. Usually he catches him within a short distance, the pathetic human merely crumbles to the ground, and lion drags trainer all over the area by pulling at his clothing, arms, or legs. Sometimes, however, the trainer eludes the cat and manages to put a considerable distance between himself and the pursuer. The lion then smiles in utter delight, crouches low to the ground, twitches his long tail once or twice, and charges. In giant bounds, as fast and as big as a freight train, the lion closes the distance. The trainer deftly deflects the onslaught by raising his forearm as a target and directs the animal's motion upward—and then collapses beneath the weight of the four-hundred-pound cat.

Amazingly enough, there's no great risk in the game, even though our four-hundred- or five-hundred-pound cats have always been *fully armed.* They have never been declawed or altered in any way. Over many months of patient training the cats have been taught never to bite hard and never to use their claws. A complete understanding, a wonderful state of rapport, has been established between man and animal and it's rather an exciting thing to see. Above all, it's a wonderful thing to experience—to see the huge cat charging at you with all that strength and power, to have him drag you down, weight you down, pull at you, even take your neck or face in that great mouth, and know, deep down inside, that it's all in fun, that he really isn't going to harm you.

I've never suffered so much as a serious scratch from any of my predator friends during the past twelve years. Yet, despite the fact that we've been successful in our affection-training endeavors, I'm utterly convinced that these big predators just do not belong in the home. We've learned from experience that the time always comes when the animals, much as we loved them, are better off in the zoo than in our living room or backyard. Not only are they happier, but we poor weak humans are infinitely more relaxed and at ease when we can handle them, play with them, and exercise them under the controlled conditions that the zoo affords.

One of the major problems in keeping big predators in a home is that all too often their owners just don't realize that lions, tigers, leopards, and other such felines have a penchant for little children. They are fascinated by any creature that approximates their own height and have an inherent obsession to chase and capture such little beings. Hediger and other animal psychologists agree that the cat merely wants to play, but that's where the danger lies. Once the young lion overtakes the child and begins to play, the child becomes frightened and attempts to escape. Instinctively, the cat's claws come out, his teeth are put to use, and the child is often badly mauled or killed.

A number of such incidents have occurred in this country in just the past few years. Some "pet" owners attempt to solve the problem by declawing their cats and also extracting the long canine teeth. Not only are such measures cruel, they are also inadequate. A two-year-old lion, deprived of its claws and canine teeth, is still quite capable of killing a human with its paw if it feels the need to do so.

Even more sadly, when the owner does realize that his crippled animal is becoming too big to curl up on the couch and that the cost of feeding him eighteen pounds of food each day is a bit too much, he will probably find that no zoo will take the animal. The lion, the zoo director will inform him, just can't be placed in company with other members of his species because, lacking claws and teeth, he can't protect himself. Quite often the only possible answer is to put the poor feline to sleep. All because his owner wished to be a little different and inflate his ego, even if it meant crippling a beautiful animal so he could keep it as a house pet.

Professional animal trainers are often guilty of even greater sins against the animal world. Trainer Pat Derby delves into the matter in depth and disgust in her fine book *The Lady and Her Tiger*. The facilities she describes, that many of the animals you see in motion pictures and on TV are kept in, make our small zoo cages seem almost palatial by comparison. She deplores the way the trainers house their animals, feed them, treat them, declaw them, extract their canine teeth, neuter them, and otherwise render them abject objects of pure misery. I agree with her completely and I love her for writing her remarkable book.

I once asked a friend of mine who trains many animals for TV serials and commercials why he found it necessary to declaw his big cats. He gave me two reasons: First, the animals often become nervous in front of the TV cameras and there is a greater possibility of the cat injuring someone if it still possesses its claws. Secondly, the leopards, lions, tigers, or whatever cat is being used, very often tear the costumes of the human actors they work with unless their claws are extracted. He confessed, too, that he really hadn't had much success in trying to teach his animals not to use their claws, if they are permitted to retain them. Apparently too few professional trainers have, according to Pat Derby.

CHAPTER 21

Zoos, a Last Hope

The zoo was closed for the winter and it was cold and snowing outside. I had just returned from the other winter building where I had enjoyed a long visit with Leonardo Lion, Bagheera Leopard, and the jaguars. I was standing in the office, my backside against the space heater, trying to get warm again.

It was a miserable day, but I was quite pleased about something for a change. Mr. VOB had scored a major victory over his budget problems and our cats, the chimp, the elephant, and the smaller monkeys now had brand-new cages, much bigger ones that permitted the animals a lot more room for exercise in their winter quarters.

This long overdue improvement was brought about by increasing public and federal governmental concern about the welfare of exotic animals in zoos. The United States Department of Agriculture was working in conjunction with the American Association of Zoological Parks and Aquariums on minimum space standards for captive animals and Dr. Gordon Hubbell, a member of the AAZPA committee, kindly sent me a copy of the space requirement proposals.

The public's concern about the zoo animal's well-being was roaringly expressed by a number of national humane groups. Certainly I, and most other zoo directors, had long known that this would happen. Many zoo directors, including myself, agreed with some of their statements, but when they stubbornly insisted that *all* zoos big or small, should be closed because zoos were "cruel" to animals, they overstated their case to the point of being ridiculous.

I confess to taking somewhat sneaky advantage of the situation. I gathered up my written material on space requirements and the news releases from the humane groups and called upon Mr. VOB in his office one day. After glancing over the material I had placed on his desk, he agreed that it might be a good time to do something about our winter quarters. At his suggestion I drew up some basic plans, and work began immediately.

While I was pleased about this happy result, I was all too aware of the fact that public sentiment was strongly against zoos. Many influential groups and people, in government and public life, were becoming increasingly loud in their demands that all zoos *immediately* provide adequate, even elaborate, facilities for the animals, or be closed.

It was rather a sad situation, not only for our little zoo, but probably for at least 80 percent of the zoos in this country. Zoos, after all, are the most popular single spectator attraction in the United States. Last year zoos drew more than 113 million visitors—more than all the pro football, basketball, baseball, and hockey games combined. Zoos appeal to everyone, regardless of age, whether the individual has a Ph.D. or is illiterate. They provide family entertainment at admission prices so low they're almost the only activity a family can enjoy together.

Zoos are of great educational value as well, and most zoos have developed educational programs that reach into the local school systems and other community areas. They also provide facilities for professional people interested in animal psychology and behavior.

Most importantly, zoos are the last hope for thousands of animal species. Just recently I talked to Gary Clarke, director of the Topeka (Kansas) Zoo, after he had returned from India. Gary had been a guest of the Indian government and had helped them, in an advisory capacity, in their attempts to save the Bengal tiger from extinction. He told me that while they were doing a much better job than the African countries in controlling poaching, the great concern in India was that the preserve areas were relatively small. He estimated that there were, at the most, some two thousand Bengal tigers left in the wild. Inasmuch as statistics state that the populations of Asia, Africa, and South America are mathematically bound to double within the next twenty years, neither of us could conclude that the outlook for the survival of the Bengal tiger was anything but dismal.

Anyone who reads the newspapers must certainly be aware that there are some fifteen thousand elephants being poached every year in Kenya alone; that less than one thousand elephants are probably alive in Uganda today; that, regardless of international treaties prohibiting the exportation or importation of thousands of species, poaching still continues at a disastrous rate. Leopards, lions, elephants, jaguars, Siberian tigers, dik-diks, warthogs—you name it—are all being slaughtered for their skins, or various parts of their anatomy are being used to make curios for the tourists.

As I've stated before, the preserve sanctuary concept that was to perpetuate all animal species in Africa, Asia, and North and South America cannot be the only and ultimate answer to the preservation of those animals whose existence is threatened. Time, poaching, agriculture, expanding civilization and cities, population explosions, droughts, and greedy politics have certainly proven this. In the future, zoos must bear a large part of the responsibility for saving these animals from complete extinction.

Yet a large proportion of zoos in this country and elsewhere are not prepared to assume this responsibility in an adequate way. And the great question is, of course, why aren't they prepared? Just what errors, what omissions, what faults contributed to the sad fact that most of the zoos in this country are years behind time?

A few years ago, an article in *Sports Illustrated* suggested that the responsibility for the sad conditions in most American zoos lay in the ultraconservative attitudes of the zoo directors. They compared the then very popular drive-through, safari-type animal parks with the cages in most of our zoos and suggested that the drive-throughs were the only and ultimate answers. As usually happens with categorical statements, this was proven in error. The majority of the drive-through parks have since gone into bankruptcy, largely because their attendance figures declined once the novelty wore off.

More recently Peter Batten, in his 1976 book *Living Trophies*, also attempted to place the blame for zoo deficiencies entirely on the tired shoulders of the zoo directors. Mr. Batten visited a great many zoos in this country and found little to please him. In my mind, he failed to get to the heart of the matter.

It is my personal observation that the zoo directors and their staffs are most certainly not to blame for the inadequacies that exist in our zoos today. I've learned to know and respect many of the directors personally and I'm a fervent believer in the ability and dedication of the directors, curators, and keepers who staff the zoos.

Needless to say, though, Batten found the Des Moines Zoo lacking in many respects, and even went out of his way to state that I was a most inadequate and incapable director. He utterly missed the main problem, in my view, but a number of influential people in this community believed his story and demanded that the Des Moines Zoo be closed. Like Batten, they too forgot that a zoo director is not an entity unto himself.

The great problem that has existed over the years is that the professional people of the zoo world have had too little to say about the administration of their zoos. In actual practice, most of the zoo directors I know are regarded as mentally just one step above the chimpanzees by the administering bodies that govern their zoos. Many zoo directors might as well be given simple cages to conduct their business from rather than offices. This would truly be more in keeping with the role they play in the planning, the policy making, and the day-to-day administration of their zoos.

Approximately two thirds of the nation's zoos are administered by municipalities or, more directly, by parks departments. A city parks department is also entrusted with the supervision of public parks, golf courses, cemeteries, swimming pools, recreation centers and programs, botanical gardens, and many other concerns. Most cities today are experiencing financial problems, and the parks department director often feels that his limited funds must be distributed as equally as possible among his many projects.

Zoos, however, have special problems. To build an adequate animal collection and proper physical facilities usually requires a large sum of money—often much more than a parks director considers a fair share of the budget. Zoos, too, are in a perpetual state of contingency. Animals become ill suddenly; they must often be purchased or sold within a matter of minutes; public relations and educational programs are constantly changing.

One of the officers of a large zoo-consultant firm insists that zoo associations or Friends-of-the-Zoo organizations do a much better job of administering zoos than municipalities do. Perhaps this is so, but here again, all too often the director and the zoo staff have very little to say, even about the day-to-day operation of the zoo, let alone about budgeting and policy-making decisions.

A few months ago I received a visit from a zoo director in our neighboring state of Illinois. As always when zoo directors get together, we compared notes on animals, animal care procedures, and personal problems. In this account of our conversation, I've changed the name of my friend, and our allusions to other zoos are generalized to prevent any possible problems that might arise for the directors if their names or the names of their zoos were mentioned.

We were seated in my little office after taking a trip of our facilities and admiring the animals. Jerry, as I shall name my visitor, pulled out a cigarette and lit it. "I can't understand why anyone would want to close your zoo," he stated unbelievingly. "You've got some nice facilities and you've got room to expand as well. What's really the problem?"

"Basically it's the same problem most of us are facing," I replied. "People want the predator animals out of cages. Some of them don't want to wait and give us the time to do just that. They want to close us up now—partly because they don't believe we can raise the money to build natural habitat exhibits."

"Well, that's not so expensive with some of the new exhibit techniques," Jerry responded.

"Of course it's not, Jerry," I agreed. "I've taken some basic concept drawings to five other zoo directors and their staffs, and they all agree that the ideas are feasible and practical. Probably the biggest problem we have here, though, is just who is going to run the zoo. The Zoo Association has openly stated that they wish to administer the zoo, and the Parks Department has definitely declared that they're not about to entrust a city enterprise to the Zoo Association."

"Oh oh," Jerry exclaimed, "I think I've heard that song many, many times before. It seems to have happened at virtually every zoo in the country at one time or another."

"Of course it has. And as you know, it's always the zoo director who gets caught in the crunch. The Zoo Association has been after my head for more than two years—ever since we obtained the camel."

"What do you mean?" Jerry asked.

"The Des Moines Kennel Club wanted to donate an animal to the zoo and we had to get it quick so they could use it to promote their dog show. That's where the money was to come from. We had a black leopard lined up first. Our governor, Bob Ray, even interceded with the U.S. Department of the Interior and secured permission to have a Texas zoo ship us the animal before we received the endangered-species permit. That wasn't good enough for the secretary of the Texas zoo's Zoo Association. He wouldn't even take the Governor of Iowa's word that it was a legitimate deal. He wanted his money first. He demanded that we slip the money under the table, so to speak, to an animal dealer in Texas, who would then slip it to him. Only then would he send the leopard. Of course it was as illegal as could be, so we turned it down." I paused for breath and continued. "We tried two other zoos for different animals, but both had to wait until their zoological associations had their monthly meetings before the poor directors could get permission to sell us the animals, so we were out of luck there, too. Finally we found a nice female camel that the dealer would deliver free of charge. That alone saved us seven hundred and fifty dollars. The Kennel Club members were delighted. My Zoo Association, however, wanted to take over the zoo—and the animal purchases—so they decided they didn't like the camel."

"Then they tried to get you fired?" Jerry asked.

"They did worse than that. They froze all the money in my special animal fund that they had promised they would never touch. The Theta sorority group here in town had given me quite a bit of money and we had agreed that it was to be used for a new cat exhibit—one like the Indianapolis tiger exhibit."

"Yes, I know what you mean. It's certainly one of the finest, inexpensive exhibits in the country. Even Peter Batten liked it."

"Well," I went on, "their next step was to convince the parks director—who has since left here—that I wouldn't cooperate with them and they demanded that the purchase of all animals be completely curtailed until a master plan was completed. That was four years ago and right now I'd be hard put to fill the cages in the exhibit area, we're so low on animals. You probably noticed that a lot of our species are singles, not pairs or family groups."

"Yep, I noticed," he replied. "I've got the same trouble at my zoo. One of my superiors absolutely insists that we keep single animals and not pairs. He thinks we can exhibit more animals that way."

I snorted in sheer disgust. "Everyone, it seems, knows how to run a zoo except the stupid directors."

"It's certainly that way at our place," Jerry said. "I'm the fifth director they've had in two years."

"Good Lord," I groaned, "that's incredible. By the way, has the vacancy been filled in that zoo across the river yet?"

"No, it's been open for more than a year now."

"I confess I was interested in it," I said, "but the director who retired told me it was a bad, bad place. The parks director is a real tyrant, he told me."

"I hear that Jim, over at the zoo east of me, is beginning to have problems."

"I've heard the same story," I answered. "Seems kind of funny in a way. He left that Missouri zoo because of trouble with the Parks Department. He found a nice guy to work with in the Illinois zoo, but now the Zoo Association is trying to take it over."

"Most of us are in the same boat," Jerry said. "And now some of the humane groups are trying to create a split between the directors and the keepers. Batten seems to be trying the same thing."

I nodded in agreement. Then I asked the one question everyone seems to get to, sooner or later. "How much are they paying you?"

"Eleven thousand," he replied. "My wife works as my secretary. They pay her the minimum, two dollars and thirty cents an hour, for that."

Jerry had formerly been one of the top curators in one of the largest zoos in the country. He had quit that job to take one with a meat-processing company at $16,000 a year. He returned to the zoo world simply because he missed the animals so much.

"Figuring that eleven thousand on an hourly basis, that means you're getting less than your keepers, doesn't it?" I asked.

"Oh, sure," he answered. "After all, you know as well as I do that we have to be on the job virtually every day."

"Right," I exclaimed. "I've had forty-six days off, total, in ten years, and that includes Saturdays and Sundays. I've had five days vacation, and I've spent those at other zoos trying to get advice on our own problems. I keep a diary on my time and activities every day and I've found I have never put in less than thirty-five hundred hours a year on zoo activities. That puts me on a salary of about four dollars an hour. My regular keepers make about a dollar an hour more than I do."

Jerry looked at me for a long minute. "Do you suppose we're crazy or something?" he asked.

"That's just what my wife and some of my friends want to know."

"I suppose we do it because we like the animals," he said thoughtfully. "I can't think of any other reason, that's for sure."

"I think you're right," I said. "What else is there but the animals?"

Jerry left a few minutes later to return to his zoo and his problems. I sat there thinking for a long, long time. Just how in the heck, I asked myself, could I convince a lot of people who wouldn't even try to understand that a zoo is different. That zoos are a constantly changing reality and don't always fit into the rules of predictable logic, as some people insist that they must. How could I explain to them, I wondered, that unless the expertise of the zoo's director and staff are listened to and considered, the results are simply chaotic. Pretty soon, I muttered half-aloud, a lot of the zoos in this country are going to close their gates forever if the people that administer them don't start listening to the directors and the zoo staffs.

Then please don't suggest just sending the animals back to Asia, Africa, or South America. All zoo directors have heard that too often.

They're exterminating the animals in those areas, in one way or another. They don't want them, apparently, unless their pelts are in prime condition and will bring a good price on the poachers' market.

A few days after Jerry's visit to our zoo the Des Moines book stores began displaying copies of the book by Peter Batten we had talked about.

Jerry, like his many predecessors, has since resigned from his Illinois zoo. Batten's book, *Living Trophies*, however, is still very much with us. Its influence hangs like a black cloud over our little zoo. Our local newspaper printed an editorial endorsing Batten's criticism of our zoo, and immediately many others joined the influential ones in clamoring for improvement, if not the absolute closing of the gates.

I'll never forget one incident in particular. It was quite typical of the criticism and my tired attempts to explain that Batten was not the voice of the Almighty. I had just finished giving elephant rides for two long hours. The thermometer registered ninety-six degrees, and I was exhausted and about to have a heat stroke. I was leaning against the bars of the elephant enclosure, struggling for breath, soaked with perspiration.

A man approached. He was well attired, in his forties, and appeared intelligent. "You look warm," he suggested.

I glanced up at him and shook my head in sad agreement. "Yes," I replied, "I am a little warm."

He hesitated a moment, then asked, "Why do you do all these things? Is it part of your job to give elephant rides? Do the people in city hall demand that you give elephant rides?"

"No, not at all," I assured him. "We do it for the same reason other zoos do. Children love it and it increases our attendance and revenue."

He thought about that for a moment. "And after this," he continued, "you give a reptile lecture and milk rattlesnakes and charm cobras. Is that right?"

I assured him that if I recovered in time, I would be doing just that in another half-hour.

The man shook his head "You know," he exclaimed, "I just don't get it. And I'm not sure that I approve of it at all. I've just finished reading Peter Batten's book and he states that you're more of a public relations man than a zoo director."

"Well," I replied, laughing, "he said that about a lot of other zoo directors, too, didn't he? In fact, my friend, I don't believe Mr. Batten approved of anything, hardly, about any zoo or zoo director in the country. After all, every zoo has to have something of a public relations program. There's no other way of obtaining interest and support for the zoo unless you have a PR program."

"Isn't the Zoo Association supposed to do that?"

"In some zoos they do raise money for different projects. Here they haven't been particularly successful in that endeavor. We have another problem here, too. The city of Des Moines has a philosophy that the people who use our public institutions should pay a fair portion of the cost and operating budget. I think that's only fair, too. If this zoo, for instance, doesn't pay a fair portion, through admission revenue, of what it costs the taxpayers to fund it, it's going to be closed down. It's as simple as that."

"According to Batten, maybe that's just what should be done to a lot of zoos," my critic stated.

"Friend," I tried to explain, "there are over a thousand animals on the United States Department of the Interior's endangered species list right now and more are being added every day. Zoos have saved many species from complete extinction and they're faced with the responsibility of saving countless more. What this country and the world needs is more zoos—adequate and good ones, I'll admit—not fewer ones. There is, after all, a nice little mathematical equation in genetics that demands that there be a rather considerable number of animals in a gene pool, if the species is to survive at all."

"Well, I don't know anything about that," the stranger said, "but Batten certainly did say that your cages were too small."

"Batten was speaking about the cages in the winter quarters," I countered, "but that's all been corrected. The animals have plenty of room now." I wiped some of the perspiration from my face with a handkerchief. "But as far as that's concerned, I don't believe in cages for our big cats at all," I continued. "They should have larger, open facilities. That's something I've been trying to accomplish for ten long years now. Maybe some day we'll get it done. There are new exhibit techniques that are relatively inexpensive—like the ones at the Indianapolis Zoo—so I still have hope of solving that problem if people like you will just give us enough time."

"I'm not trying to be unfair," the man stammered, "but I'm not at all convinced that Des Moines can support the up-to-date, natural habitat zoo that you have in mind."

"You're not the first person to question that," I admitted. "We brought in one of the finest consulting firms in the U.S. just to answer that very question two years ago. They concluded that our greater metropolitan population would provide an attendance figure of some three hundred thousand visitors each year and could support a six-million-dollar zoo expansion. Frankly, I'm not quite that optimistic. I'd settle for a hundred-and-fifty-thousand attendance figure and if someone would give me just eight hundred thousand dollars I'm convinced that a year-round exhibit building with nice big open areas for cats could be constructed. If they can do it in Bloomington, Illinois—which is much smaller than Des Moines, incidentally—we can do it here."

It was getting time for me to prepare for the reptile lecture. I asked him if he wished to stick around and watch the program. He hesitated a minute. Really, I assured him, it did have a great deal of educational value. One narrator, I explained, had a master's degree in herpetology. He might learn something that would interest him.

"Maybe so," he agreed, "but I think most of the spectators will be sitting there watching in the hope that one of the cobras will bite you and they'll see some excitement."

"There's possibly a little of that, too," I admitted, "but essentially I think they're really interested in the cobra itself. Very few people have seen cobra charming in this country and I suppose they feel something of the fascination humans have for poisonous snakes all over the world."

A thought occurred to me and I had to laugh. "Perhaps you're thinking of the statement Batten made in his book about me being an amateur snake charmer. Really, I'm not that bad. In this business you're either very good or you're very dead. I've charmed these cobras hundreds and hundreds of times and I'm still quite alive. I confess I cheat a little, though. I have plenty of immunity to counteract any toxin the cobras could give me."

The man admitted that there might be something to that. He stayed for the lecture and looked as though he enjoyed himself. I'm not convinced that I sold him on the value or necessity of zoos, elephant rides, reptile lectures, and cobra charming, but I do believe I gave him something to think about.

CHAPTER 22

Free From Fear

One afternoon I noticed two of our girl guides standing in front of the tiger's cage. The great cat was lying on the big concrete shelf that forms the top of his den. He was gazing at something in the distance, head high, intent and motionless. The girls, equally absorbed, were watching the tiger. For a very long time they stood there, without talking or moving, almost as though they were hypnotized. Then they moved on and went about their duties.

I smiled to myself. Over the years I had seen so many people do much the same thing, whether they were casual visitors or members of a tour group. The big cats, the wolves, and the eagles seemed to captivate individuals and hold them spellbound.

I knew what the girl guides and the others had seen and what had kept them there for such a long time. They were entranced by the sheer beauty, the size, the alertness, and hauteur of the predators. They too, I was certain, had felt much the same wonder and awe I had experienced that day many years before when I first held a peregrine falcon on my gloved hand, cast her free, and watched her circle, luff into the wind, and rise into the sky. For long minutes she hovered a thousand feet above me, wings pulsing, holding her position against the wind. Then I threw out the lure and the falcon folded her wings and plummeted earthward, sliding down the sky in a breathtaking, two-hundred-mile-per-hour stoop. Nearing the lure, she leveled off and landed gracefully on it.

On that day a window opened into a new dimension of my life and I saw, for the first time, a world of beauty, grace, and power that has enthralled me completely ever since.

I was a young man then, and since that time I've talked to other falconers, animal lovers, and research students who have watched and studied animals and we all seem to share the same feeling. We understand each other.

Yet I often wonder what it is we really see. What do the animals really mean to us? When we walk our tigers in the evening, I've watched them blend into the weeds and brush and become almost invisible, even mystical, as they turn to face us with their dark stripes and the white patches above their eyes. I have spent hours, too, contemplating the chiseled, rocklike perfection of a lion's head and marveling at the perfect symmetry of the leopards and the quicksilver motion of a cobra on the ground. To me this is beauty: beauty etched by eons of time into perfect proportion, balance, and motion. This is the superb beauty of life that, infinitely more than human art, captures the vastness of space and the unending reach of time and being.

I believe I see the beauty and innate majesty of the big cats, the wolf, and the eagle in much the same way as the ancient Egyptians saw the animals of their world. They saw them, with an inner eye, moving down endless corridors of time, in a flowing river of becoming and being, without beginning, without end. They identified with them as brother beings. The falcon was the eye of the spirit, and it became the Ka falcon, the symbol of power. The falcon, too, was the eye of the soul and it became the Ba falcon, with a human head, the symbol of man's infinite past and future and his close, everlasting kinship with his fellow animals.

Even today, I believe, we see the great predators as symbols of power: the power that overcomes, conquers, and endures, and we still worship it. We sense that such power transcends challenge, even death, and in our great fear, we primates worship it wrongly quite often and idolize the physical strength of the gladiator, the cleverness of the thief, the destructiveness of the paid assassin.

We see and sense this power of the predator with our intellect and instinctive fears, and our primate emotions project absolute thoughts of killer cobras, man-eating lions, tigers, and wolves, rogue elephants, even monster gorillas. We have been taught to think of them in this way. Yet, as Iain and Oria Douglas-Hamilton stated in their fine book *Among the Elephants*, to really know an animal one must raise it and live with it from infancy.

I have been wonderfully fortunate in this respect, for I have lived and grown with many big cats, wolves, eagles, and other predators. I have learned to see them with my heart as well as with my intellect and instincts. I have learned to recognize the shyness and fear behind their aggression, their need for warmth and companionship. I have played with lions, tigers, leopards, and wolves and learned to appreciate their gentleness, their high intelligence, even their roguish sense of humor.

On Easter Sunday evening in 1976 I found, in an unforgettable manner, just how deeply I had learned to know and love them. We were finishing dinner around 7:00 P.M. when my son, Joel, burst into the room with very disturbing news.

"We've got trouble, Dad," he shouted. "Brucie Tiger is out of his cage."

My heart stopped beating for a minute, I'm sure. We did have trouble, indeed. Brucie, our famous honorary watchcat of years before, had been confined to his cage for the past two years and had not been out on a leash during that time. We had been overwhelmed with the task of affection-training a myriad of younger cats that had been born at the zoo. In fact, I hadn't been near Brucie Tiger's cage for more than two weeks. I had spent most of that time in the hospital while the good doctors gradually lowered my blood pressure to a reasonable level.

We immediately proceeded to load the tranquilizer pistol and the tranquilizer rifle. In the rifle we placed a lethal ten-cc dose of nicotine toxin that would kill the tiger in just a few minutes. We loaded the small pistol with a normal dose of a sedative that would put Brucie to sleep in twenty or thirty minutes. As an added precaution, we loaded a shotgun, a heavy rifle, and an automatic pistol.

My wife called the police while we prepared, told them the situation, and asked them to stand by to help in any way they could. She also suggested they surround the entire zoo area with squad cars to prevent the big cat from escaping into the surrounding neighborhood.

We were as prepared as we could be for a very dangerous situation. The last thing in the world I wanted was to destroy our beautiful tiger, but I was forced to admit to myself that when big cats get out of their cages and begin roaming about, they may become frightened, terribly aggressive, and almost impossible to recapture without killing them. Just such an incident had taken place two weeks before when two leopards escaped from a zoo in the western part of the United States. Both cats had to be shot and killed.

I also recalled the story Wolfgang Holzmair had told me about a lion escape from the circus compound in Florida. A group of men went after the cat with an arsenal of weapons. The lion still managed to kill one of the men before he was downed.

Rob, Joel, Becky, and I gathered our equipment together, jumped into the car, and drove to the zoo's exhibit area. All I could think about on the way down was another recent incident when a zoo vet was mauled attempting to get a tiger back into his cage after the animal had escaped. The thought did little to lower my blood pressure.

We opened the main gate to the exhibit area (and left it open for the policemen to enter when they arrived) and drove slowly around the zoo toward Brucie's cage. We had no trouble finding him. The car's headlights picked up his eyes, burning like two fiery coals, in the corner of the exercise area just outside his cage.

"That's funny," Joel commented, "that's just where he was when I spotted him before."

We drove closer, cautiously, until the headlights were directly on him. I stopped the car just short of the four-foot-high barrier fence that bordered the front of the exercise area. Behind him and to the right was a six-foot fence. Neither would make the slightest difference to a cat that could scale a sixteen-foot-high fence without exerting himself.

My heart was pounding, my throat so dry with fear I could hardly talk. For heaven's sake, I told myself, calm down and get control of yourself. Try to think. Don't panic; just try to think and reason this out somehow.

We had two choices. I could shoot the tiger with the lethal dose of nicotine in the tranquilizer rifle and everything would be over in just a minute or two. And my tiger would be dead. Or I could hit him with the regular tranquilizer drug and he might do anything and everything in the twenty minutes or so before he went to sleep.

In any event, I realized I would have to get out of the car because the front barrier fence made it impossible to hit him from inside the vehicle.

In the meantime, while we were discussing all the possibilities as rapidly as possible, Brucie Tiger remained just where he was, apparently as calm and collected as could be, despite the fact that the car's lights were still turned on him.

"I've got to get out of the car," I told my kids. "There's no way of hitting him from here, shooting from the window."

"Dad," Rob cautioned me, "he's only about fifty feet away. If he charges, he'll come like lighting. Can you get back inside in time?"

"I don't know," I confessed. "What's more, I don't know whether the glass will even begin to stop him if he really wants to get us and hits it hard."

Whatever we did, I realized that I also had the responsibility of protecting my two sons and my daughter. And I had learned over the years that the noise of the tranquilizer gun and the impact of the big dart into the animal's body always provoked an instant, angry charge from the animal in the cage.

In any event, I told myself again, you've got to get out of the car. Taking the rifle loaded with the deadly nicotine in my hand, I opened the door and stepped out. I was prepared for the worst.

"Whatever happens, you kids stay in the car," I told them.

"No way, Dad," Rob replied. "If he charges we're coming out with the rifle and the shotgun."

I managed a sickly smile. "If he gets me down I don't think you'll have much time to do anything that would help me."

The moment I walked away from the car Brucie rose to his feet, stretched lazily, and greeted me with his friendly "pfffff." I knew instantly and surely that everything would be all right. Brucie was happy to see me. He wasn't afraid.

I returned to the car, exchanged the rifle for the pistol with the normal sedative dose, and again approached the barrier fence. Rob and Joel followed a few feet behind, each holding a rifle or shotgun.

Brucie Tiger walked casually over to the fence to greet me with his little happy sound.

I returned his "pfffff" greeting immediately and the great animal reached the barrier, rubbed against it, and permitted me to stroke his head and his back. Then, rising on his hind feet, he placed his front paws on my shoulder. His weight pushed me backward and then, to my consternation, he crouched and prepared to jump over the barrier fence to get closer to me. This was the last thing in the world I wanted him to do. My two sons and Becky were only a few feet away and I was afraid he might attack them if he got on the other side of the fence.

I distracted him by moving along the length of the barrier. "Come, come, Brucie," I called to him. "Come, come." It was the old command we used when we had taken him out on the leash years ago. He heard it, remembered it, and obeyed. He followed me down the fence on the other side, but when I stopped he again rose up and placed his paws on my shoulders. Suddenly, I had an idea.

"I think we can trust him," I told the others. "I think that if I can just get over this fence I can lead him back into his cage."

"Oh God, Dad," Becky cried, "don't do anything like that. If he wrestles you down, you'll never be able to get up again. He's so big he'll kill you just playing with you. Use the tranquilizer pistol."

She was right. With my crippled hand I couldn't vault the fence in time to keep Brucie from jumping all over me. Then too, Brucie was no longer wearing a collar, and without that, and the leash to lead him, I would have a minimum of control over his movements. The only reasonable solution was to tranquilize him.

I knew I couldn't shoot Brucie from two feet away—where he was now seated, head cocked to one side, watching me curiously. If the noise and the impact frightened him and he charged, the little fence wouldn't deter him for a second. He could kill me instantly if he chose.

"I've got to get him farther away from me, somehow," I told the others. "At least enough distance for me to have a chance at getting back into the car if he charges."

Becky came up with the solution. "Throw the leather gun case into the center of the area, Dad," she suggested. "Then while he's playing with that, maybe you can get a better shot at him."

I tossed the small carrying case into the open area and Brucie bounded happily after it. While he crouched, chewing on the leather, I lifted the tranquilizer pistol and prepared to fire.

My hand was shaking. I steadied it with the other one. I knew that if with God's help I should succeed in hitting him and he didn't charge, I would probably be the only person in history to shoot a full-grown tiger, in the open, with a dart gun and live to tell about it. The thought was of no comfort whatsoever.

Yet somehow, deep inside, I felt I could trust Brucie. I wanted him to live so much I decided to take the risk.

I placed my hands on the guard fence, lined up the sights, prayed, and pulled the trigger. The big five-inch dart struck him perfectly in the hip and penetrated deep into the muscle. Instead of an instant, snarling charge to kill me, Brucie merely looked up, made his friendly "pfffff" sound at me, and continued to play with the gun case. I crept back into the car. I was soaked with perspiration and weak from fear. Rob and Joel followed me back inside.

Twenty minutes later, Brucie was completely sedated. We called to the two waiting policemen who had been standing about a hundred yards away. They held their shotguns on Brucie, just as a precaution, while we placed him on the zoo's animal stretcher and carried him back into his cage. We gave him an injection of atropine to control his salivation and some Valium to prevent convulsions. Then the four of us sat down and waited, watching him closely for any signs of distress. Brucie slept as quietly as an old log.

I thanked the policemen for standing by and they departed.

While we waited we asked ourselves a lot of questions. There was no doubt that Brucie had escaped because of "keeper error." Two of the zoo keepers had cleaned Brucie's cage that morning. There was not enough water pressure to wash it adequately from the outside, so they had locked Brucie in his den, entered the cage, and hosed out the heavy excreta from inside the cage. They had then left the cage, forgetting to latch and lock the main door, and also forgetting to release Brucie from his den.

That afternoon Joel had approached the cage with the tiger's food and opened the den door so the tiger could come out and eat. The main cage door was closed, and he failed to notice that it was not properly secured.

"What I don't understand," Becky exclaimed, "is why Brucie just stayed in his exercise area and didn't roam around the zoo or the entire south side of Des Moines. Apparently," she continued, "he didn't even bother the two fallow deer just on the other side of the fence."

Rob is a psych major in college. He came up with a possible explanation "I think the tiger was following a pattern," he explained. "He was lonesome and he wanted to go for a walk on the leash and play with Dad like he did two years ago. He was just sitting there waiting for Dad to come along and play with him."

"Well, he certainly gets pretty lonesome, there's no doubt about that," Joel agreed. "He starts to cry for attention the moment Dad enters the exhibit area and Brucie spots him. And that's quite a distance from here."

"What I find difficult to believe, though, is that Brucie remembered his training and evidenced so much affection, considering that I hadn't worked with him for two years," I said. "I haven't even seen him for two weeks. And why he didn't attack or even threaten me when the tranquilizer dart hit is something I'll never understand."

We asked questions we'll never be able to answer with certainty, but one thread of continuity lent a possible explanation to the tiger's behavior. Essentially, Brucie was not a frightened tiger; therefore he was not a dangerous tiger. He had stayed in his exercise area quite happily, and he had not attacked from fear that we might injure or threaten him in any way. Our endless hours of affection-training had paid off. Brucie was free from fear.

Now that all the danger was past I broke down. I suppose it was a release from the terrible tension and perhaps a deep happiness that Brucie hadn't killed anyone or hadn't had to be destroyed, but suddenly I was so overcome with emotion I couldn't speak. Tears filled my eyes and ran down my face. I was glad it was too dark for the others to see.

As I held Brucie's great, beautiful head in my lap, waiting for him to recover, a portion of a poem by Blake came to my mind:

Tiger! Tiger! burning bright
In the forests of the night,
What immortal hand or eye
Dare shape thy fearful symmetry?

And I knew that in the fearful symmetry of the great predators, the "monsters" of our world, there also existed a rare beauty, patience, gentleness, understanding, and love.

I stroked his thick fur, feeling the heavy muscles beneath, and I realized how very, very deeply I loved this magnificent tiger and every other animal in our little zoo. I knew then, too, that I would never, could never, leave them until somehow they were freed from their cages and roaming, playing, freer in spirit, in larger open areas.

For I had learned, that night, that man and animals share many of the same feelings of loneliness, fear, and love. I had learned that the tiger is truly my brother.

CHAPTER 23

A Happy Ending?

This is a story that may have a happy ending.

During the summer of 1976 our zoo met and surmounted the biggest challenge to its existence it may ever have to face. The state legislature had passed a bill limiting property taxes throughout the state. This, as I had long feared might happen, placed a severe limitation on the Des Moines city budget. Our city manager reluctantly concluded that many things which were not absolutely necessary to the community would have to be either cut back or eliminated. He mentioned that the Science Center, the Art Center, the swimming pools, and the zoo would be the most likely to be affected.

We were permitted to open the zoo while the city council pondered the problem, but I was not able to secure a sponsor for our Saturday television show because of the dilemma. Without this free bit of public relations, the month of June was simply a catastrophe. Our attendance during that most important month was ten thousand visitors behind that of June 1975.

Then, late in June, the city council concluded that if we were to remain open the zoo would have to contribute more to its own support. They decided that admission prices would remain the same on weekdays but would be doubled on Saturdays and Sundays. Everyone on the zoo staff knew that, with our small collection of animals, our survival would now depend essentially on our special activities.

The radio and television people rallied to our support. Each and every day we gave free elephant, burro, and camel rides, charmed cobras, gave reptile lectures, and performed demonstrations with our affection-trained animals until we were virtually exhausted. Our good governor, Robert Ray, proclaimed a state zoo week and this helped immensely. One radio station broadcast a remote from the zoo. Thousands of people responded and came out for the occasion.

At the end of the season, despite a terribly hot, dry summer, we had made up our ten-thousand-attendance loss in June and even surpassed the previous year's attendance record by five thousand. Our attendance revenue was up a big, happy 25 percent over the 1975 season.

We had proven one thing beyond question: people were interested in the Des Moines Zoo and would support it, even when the admission prices were high.

Another happy thing has occurred as well. Ted Meredith, a loyal friend of the zoo, presented the Parks Department with a check for ten thousand dollars to be used to develop a master plan for the expansion and improvement of the zoo facilities. Bill Foley, our new Parks Director, has been responsible for this project.

Increasingly, after all the desperate years of struggle and controversy, it's beginning to appear that the Des Moines Zoo will, at long last, have a year-round exhibit building and big open areas for our animals to live and roam about in with at least some degree of freedom.

That day *is* approaching, I'm sure of it. And when it arrives I'm certain the lions, tigers, leopards, and wolves will, all at once and all together, give a huge roar of approval and thanks to the people of Des Moines for freeing them from their cages.

Be prepared, wherever you are, for that tremendous noise. We want everyone to know, all over the country, that the little Des Moines Zoo has at last arrived—that it is now a real zoo in the best and fullest sense of the word.